W9-CRY-393

Global Metropolitan

Global Metropolitan explores the connections between globalization and urbanization. Particular emphasis is placed on understanding the economic function of global cities, the political process of globalizing cities and the cultural significance of cosmopolitan cities.

This book explores the meaning of the globalizing project in cities: the maintaining, securing and increasing of urban economic competitiveness in a global world; the reimagining of the city; the rewriting of the city for both internal and external audiences; the construction of new spaces and the hosting of new events. Specific chapters look at the significance of signature architects, the hosting of the Summer Olympics and the role of the super-rich.

The main thesis of the book is that this discourse of globalizing is a major force in the restructuring of cities around the world. The book will show how a similar range of globalizing practices – including the hosting of mega-events, the siting of urban spectaculars, the representation of the city to a world of global flows – are important processes structuring urban socio-spatial transformation in the contemporary world.

John Rennie Short is Chair and Professor of the Department of Geography and Environmental Systems, University of Maryland, Baltimore County. He has published twenty-five books and numerous articles and is recognized as an international authority on the study of globalization and the city.

Questioning Cities

Edited by Gary Bridge, *University of Bristol, UK* and
Sophie Watson, *The Open University, UK*

The 'Questioning Cities' series brings together an unusual mix of urban scholars. Rather than taking a broadly economic approach, planning approach or more socio-cultural approach, it aims to include titles from a multi-disciplinary field of those interested in critical urban analysis. The series thus includes authors who draw on contemporary social, urban and critical theory to explore different aspects of the city. It is not therefore a series made up of books which are largely case studies of different cities and predominantly descriptive. It seeks instead to extend current debates, through in most cases excellent empirical work, and to develop sophisticated under-standings of the city from a number of disciplines including geography, sociology, politics, planning, cultural studies, philosophy and literature. The series also aims to be thoroughly international where possible, to be innovative, to surprise, and to challenge received wisdom in urban studies. Overall it will encourage a multi-disciplinary and international dialogue, always bearing in mind that simple description or empirical observation which is not located within a broader theoretical framework would not – for this series at least – be enough.

Forthcoming titles

Reason in the City of Difference
Gary Bridge

Ordinary Cities
Between modernity and development
Jenny Robinson

Global Metropolitan

Globalizing cities in a capitalist world

John Rennie Short

LONDON AND NEW YORK

First published 2004
by Routledge
11 New Fetter Lane, London EC4P 4EE

Simultaneously published in the USA and Canada
by Routledge
29 West 35th Street, New York, NY 10001

Routledge is an imprint of the Taylor & Francis Group

© 2004 John Rennie Short

Typeset in Times and Bauhaus by Exe Valley Dataset Ltd, Exeter
Printed and bound in Great Britain by The Cromwell Press, Trowbridge,
Wiltshire

British Library Cataloguing in Publication Data
A catalogue record for this book is available
from the British Library

Library of Congress Cataloging in Publication Data
Short, John Rennie, 1951–
    p. cm. – (Questioning cities series)
    Includes bibliographical references.
    1. Metropolitan areas. 2. Urbanization. 3. Globalization.
    4. Cosmopolitanism. 5. City dwellers–Social conditions. 6. City
    dwellers–Economic conditions. 7. City and town life. I. Title. II. Series.
HT330.S564 2004
307.76–dc22   2003024744

ISBN 0–415–30541–1 (hbk)
ISBN 0–415–30542–x (pbk)

Contents

Illustrations

Figures

Tables

Acknowledgments

This book is the result of a series of collaborative efforts. If I use the term *we* in some of the chapters that follow, it is not an arch, stylistic convention but a statement of fact. Much of the work reported here grew directly out of discussions with others, collaborative ventures and co-published work.

The material in Chapters 3 and 4 was the result of a graduate seminar held in 2000 with Carrie Breitbach, Steve Buckman and Jamey Essex. The animated but always enjoyable discussions led to a published paper in the journal *City*.

Much of the work for this book was done while I was the Leverhulme Visiting Professor to the UK in the first six months of 2002. I spent most of the time at the Department of Geography at Loughborough University. I am grateful to the Leverhulme Trust and the Loughborough Geography Department for providing me with such a great opportunity. The whole Department made me very welcome and, in particular, Peter Taylor and his colleagues in the Globalization and World Cities (GAWC) study group provided a friendly and stimulating environment for quiet thinking and slow writing. Chapter 2 grew out of an invitation to give the GAWC annual lecture in January 2002; Chapter 5 grew out of data carefully assembled by GAWC and made freely available to me; Chapter 8 was written using the tremendous sports science material at Loughborough University and Chapter 9 grew from an initial idea by Peter Taylor and a subsequent series of drafts written by myself, Jon Beaverstock and Phil Hubbard. I owe a great debt to the GAWC team who embody the very best ideals of collaborative and open scholarship.

During my stay in the UK in 2002 I also visited a number of universities including Barcelona, Bristol, Cambridge, Durham, King's College London, Leicester, Oxford and Southampton. Responses to my lectures have informed and shaped many of the themes in this book. A large number of people were involved in organizing my visits and to each of them many thanks.

Let me acknowledge the love and support of Lisa Benton-Short for the compulsive-obsessive writer that is her husband.

1 Introduction

This book will explore the connections between globalization and urbanization. Emphasis will be placed on the economic function of global cities, the political process of globalizing cities and the cultural significance of cosmopolitan cities. Perhaps the easiest way to begin is with three, very brief, urban stories.

First: the changing face of Syracuse, New York. In the 1970s the city of Syracuse in upstate New York changed its logo. The old logo, in operation for almost 100 years, depicted an industrial city where tall factory chimneys proudly broke the skyline. However, the global shift in manufacturing employment eroded much of the city's traditional economic base. The new logo, in contrast, showed a postindustrial landscape, sleek buildings overlooking a pristine lake. The change in imagery was matched by the change in landscape. In the early 1990s a new postmodern mall, the Carousel Center, was built on the site of an abandoned petrochemical storage site in the middle of a landscape of deindustrialization and decay. A postindustrial iconography replaced industrial imagery as the basis of the city's new identity. In October 2002 ground was broken, beside the new mall, on a new $160 million hotel that was the opening stage of an eventual mega-mall, DestiNY, that is intended to cover 3.2 million square feet, cost $3.2 billion and consist of 4,000 hotel rooms, 30 restaurants, 65 acres of an indoor park as well as a miniature replica of the Erie Canal.

Second: situated on the other side of the global shift in manufacturing employment is the Chinese city of Shanghai. Since the early 1990s Shanghai has built a downtown of high-rise towers, a new international airport and under construction are a deep-water container terminal and a magnetically levitated train. Plans are on the books to build the world's tallest building. A new global city was being self-consciously recreated.

Third: in 2002 the tourist board of Rio de Janeiro was seriously considering suing the media conglomerate Fox because, in an episode of *The Simpsons* the city was described as "a city where all men are bisexual, where fearsome monkeys roamed the street, and tourists are kidnapped by taxi drivers and mugged by children." A fictional dance, the "penetrada" was also mentioned. All this, felt the city officials, caused damage to Rio's international image and loss of revenue.

While there is uniqueness in each of these tales, there is something more universal. They are part of a broader story of urban change that is sweeping the world. A global urban restructuring is taking place that, internally, involves a social and spatial restructuring, and, externally, a repositioning and re-presentation in the world of flows, images and symbols. Important socio-spatial transformations and major semiotic changes are reshaping the new global urban order.

There are some definitional issues that we need to consider before proceeding. We need to define a global city and indeed the whole notion of globalization.

DEFINING A GLOBAL CITY

Some definitions are clearly in order, not so much to put up barriers but as creative aids to some systematic theorizing. Global cities are big, but not all big cities are global cities. Tehran, for example, has a population of almost 11 million. It is not a global city. We can distinguish between *megacities*, those large and increasingly growing cities found throughout the world but especially in the Third World, including such cities as Tehran, and global cities that are also large but have other characteristics. We can also make a distinction between world cities and global cities. *World cities* are large cities linked, however loosely, to a global urban network of flows of people, goods, ideas, practices and performances. *Global cities*, in contrast, are the core of this global urban network. If we use the hierarchy metaphor, although later I will undermine it, they would be the apex; if we use the network metaphor they are the main switching points, the really connected world cities. The distinction is loose. When does a world city become a global city? The answer, unsatisfactory to many I am sure, is "It depends." If we are looking at cinema then Los Angeles clearly becomes a global city, but in terms of other economic transactions, Los Angeles is relatively provincial, despite the academic boosterism of the LA School to the contrary. In terms of strategic surveillance and military action Washington is the center of the global empire that is the United States but the city fails to make global status in terms of economic flows. We can use a soft as opposed to a hard definition of global city. Like being in love, being a global city can be a temporary phenomenon, a sought-after status, sometimes achieved but always a source of motivation and longing. Thus a megacity may become a world city if there is a significant inflow of foreign tourists. And a world city may become a global city for at least two weeks while it is hosting the Summer Olympic Games. This more plastic use of the terms undermines definitive lists of global cities. I prefer the term global cities to world cities because it is possible to use the term *globalizing city* to capture that sense of becoming and longing. Globalizing cities are both global cities seeking to maintain their position and non-global cities seeking to become global cities. The terms are not permanent unchanging verities, but relational, spectral, temporal, shifting and unstable.

We can be slightly more precise, or suggest other folds, to use a post-structuralist term, by considering both the minimum attributes of global cities as well as their threshold network function. An important ontological point is their size. A base line population of at least 1 million may be a basic prerequisite for being a global city. The figure is selected for its symbolic rather than any functional functions. Economic attributional characteristics include a concentration of advanced services, creative and cultural industries and command and control functions. Social attributional characteristics include a polyculturalism that encompasses and embraces various ethnicities and cultural diversities. Network characteristics include being an important node in the global flows of people, goods and especially ideas. Global cities are networked into circuits of transnational movements, cultural diffusions and planetary economic transactions. Global cities are the command centers of a global economy, the connecting points of a global society and important sites of social and economic transformation.

GLOBALIZATION

A lot has been written about globalization. Here I want expose some of the popular representations to closer scrutiny and unpack some of the different meanings of globalization. The three most popular conceptions of globalization are that it is a new thing, it makes everywhere the same and it is bad thing.

Globalization as something new? Although the rate of globalization has certainly increased in recent years, we have been living in a truly global world for over 500 years. When Columbus landed on the Caribbean island a global exchange between the eastern and western hemispheres created a truly global world. People, animals, viruses as well as goods and ideas now moved between formerly separate continents. Spanish was spoken in the New World, potatoes were introduced into the Old World, slaves were shipped from Africa to the Americas while gold was sent from the Americas to Spain. It is more appropriate to think of globalization as a series of pulses over the subsequent five centuries producing new configurations. These reglobalizations connect and disconnect different places in different ways. So rather than positing a sharp dichotomy of global/non-global, most of the world has witnessed various waves of globalization and reglobalization. Between 1880 and 1914, for example, an epoch marked by a significant upswing on globalization trends, a whole set of international standards ands organizations were established including the Olympic Games and various sports federations as well as postal unions and international time zone agreements.

Is globalization making everywhere the same? There are powerful homogenizing tendencies. Since 1989 a capitalist system has dominated the whole world. We are all capitalists now. There is now very little alternative to capitalism. And while there are still a few Marxists in tenured positions at

US universities, there are few competing economic ideologies that have captured popular interest. Indeed the hegemony of the neo-liberal agenda is so complete that even resistances and alternatives are situated in response to its doctrines.

The world is becoming smaller. The hero of Jules Verne's novel *Around the World in 80 Days*, first published in 1873, reads a report that it is possible to travel around the world in 80 days. While he believes the claim, none of his friends do, and so he makes a wager and thus begins the adventure of the novel. At the time of Verne's writing 80 days was an optimistic estimate. Today it is more likely to be 80 hours. In 1960 it was Marshall McLuhan who coined the term the global village to refer to a world tied together by images and media. Since McLuhan committed that phrase to print the world has shrunk, with more frequent and cheaper international travel, the Internet and the globalizing of shared images and news reports. The world has become smaller in space-time. Places are closer together. However, small differences become even more important. Small differences in wage rates between different parts of the world take on huge significance and accessibility once measured in days and months is now calibrated with reference to hours, minutes and even seconds. Global cities are places of concentrated global accessibility different from places just a few hours away. The more the world had become a global village, the more differences within the village matter.

While there are a range of similar goods and images available around the world, they are used and incorporated in different ways in different parts of the world. Different locales have different ensembles of the same images and goods. While there are shared languages of consumption and exchange, the regional variations are still important. Think of the way that multinational corporations now have strategies of glocalization in which global commodities are branded for different markets. In Ghana, Guinness is marketed as a distinctly African beverage.

Globalization has created complex patterns of hybrids rather than a common standard of homogeneity. Similar goods are consumed differently around the world. The English language has been creolized into different dialects and more people now use a number of different languages and variants of the same language depending on the context of communication. Japanese marriages will often consist of a traditional and a more Western ceremony; people will consider themselves citizens of more than one country. More groups are combining local and national identities with a cosmopolitan identity to produce a rich mosaic of different identities rather than one all-encompassing global identity. Globalization is creating more hybrid identities.

Globalization is a bad thing? The term has taken on a darker hue, often standing in for capitalism, rapid capital mobility, loss of jobs and a whole range of negative trends and processes. Globalization has become the catch-all phrase for the negative things of the contemporary world. Globalization as the root of all modern evil is a tidy if inaccurate reading. Not all forms of globalization, however, are bad; we can identify more benign global discourses of environmentalism, human rights, social justice and economic equity. There are growing senses of a global community and global standards of social justice, environmental quality and political rights. World public opinion created and maintained by global media coverage has often been an important lens in which national dictatorships and local regimes are viewed. Cosmopolitanism is now a more active agent in global politics and domestic power relations. To be sure there are limits. While Saddam Hussein was roundly condemned in the court of world public opinion for the abuses of human rights, his regime was only toppled by the brute force of superior military power. World public opinion is important, but less effective in forcing regime change than is military superiority.

Globalization is a super-condensed word on which a variety of many different meanings and interpretations have liquefied from the hot air of discussion. We can make a rough distinction between economic, political and cultural globalization. The distinction is an analytical conceit since in practice the three forms are more linked than separate.

Economic globalization involves the more rapid flows of capital around the world, the lengthening production chains of goods and services across borders and the increasing interconnectivity between the economies of different countries. During a previous round of profound globalization, Karl Marx wrote of capitalism as an agent of great social change; "everything solid melts into air" he noted in an arresting image. The present day round of global capitalism is also profoundly transforming. There has been a global shift in manufacturing and a consequent decline of the male working class in Europe and North America and the creation of a new female working class in South and East Asia. National territories have lost their spatial homogeneity as islands of global connectivity differ increasingly sharply from the rest of the national space economy. Occupational rewards differ enormously depending on the form of the global connect. Thus in the USA basketball players can earn in a year what a steel worker earns in a lifetime. One is positively connected to the global economy, the other negatively. Economic globalization is creating profound differences sectorally between economic sectors and job categories, and spatially between parts of the country and the city.

Political globalization is evident in such global discourses as trade, aid, security and environmental issues. The world is now organized along more global systems of regulation, monitoring and control. This is not to argue for the death of the nation state. In fact, distinct elements of the nation state are reinforced by globalization as some parts, especially a central banking system and trade departments, play a pivotal role in managing the global national nexus. Since 1989 a bipolar world has been replaced by a world dominated by US military power and a neo-liberal political economy. While there are differences between countries, they can be more easily situated on one global metric (one global dimension of comparison) of open markets, which in effect is the openness to capitalist penetration. There are resistances and reaction. When the present is ever changing and the future is uncertain the past becomes a place of security. The rise of fundamentalism, Islamic, Christian and Jewish, is a promised return to fixed, unchanging categories. Fundamentalism provides a sense of certainty, a soft landing for the troubled and something solid that promises not to melt into air. Fundamentalism and cosmopolitanism are the two interconnected poles of the social reactions to globalization.

Cultural globalization is the degree to which similar cultural forms are found around the world. This has led some to argue that the process of cultural homogenization, often portrayed as a form of Americanization, is occurring. While US popular cultural forms are disseminating much more wildly and deeply around the world, more people speak a form of English, eat at McDonald's and watch Hollywood movies, this has not led to an upsurge in pro American sentiment. If anything, quite the reverse. There is a subtle difference between the production of cultural forms and their consumption across the world. The consumption of culture is not a passive process of indoctrination, but a more active process of incorporation and creative readings. While there are similar cultural forms, their consumption varies across the world. If anything there is increased difference as new and old, indigenous and exotic cultural forms are tied together in new and creative ways. The process of cultural globalization is creating as much difference as similarity. New cultural identities are being created around hybrid forms as well as around invented traditions that are resistances to perceived cultural imperialism.

The current round of globalization is creating difference and divergence around the world. Minor differences in wage rates are exacerbated in a truly global economy, new hybrid identities are created as culture is prized loose from its traditional locational constraints. When Detroit is becoming an important "Arab" city then the traditional definitions of both Arab and Detroit are undermined and reshaped. Difference and divergence, change and hybridism are the new order of the day. Globalization is restructuring the traditional forms of economic, political and cultural difference and similarity.

Dominant narratives have a way of crowding out alternatives. The emphases of the dominant globalization narrative are of an integrating world economy, a homogenizing global culture and a coherent global polity. It may be instructive to end this section by noting the possibility of an alternative discourses that focus on globalization as a process that generates fractured economies, splintering cultures and resurgent nation-states.

THE BASIC ARGUMENT

While there has been great deal of research on what have been called world cities, much of this research fails to capture the processes of this global urban change. I want to consider a change of name that signifies a change of approach. Rather than using the term world city I want to use the term global city. As I stated on p. 21, I prefer the term "global" because it is possible to use it in a verbal and an adverbal as well as an adjectival form. It is indicative of cities being active agents and forces us to consider the notion that globalization is enacted and performed. This formulation begs a number of questions. First, are not cities places rather active agents? True, but we can think of both urban regimes and urban alliances as different political groupings that create compacts to make a global connection. The city is an arena for a constellation of distinct social interests that are negotiating the global–local connection. Second, what exactly is this globalizing project? Fundamentally, it is the maintaining, securing and increasing of urban economic competitiveness in a global world. This involves many things (as well as counterpoints) including, but not restricted to global connections (new nationalisms), global identities (new populisms), a self-conscious global look and feel (the invention of the local and the rise of fundamentalisms). We can identify the modalities of globalizing cities that encapsulate a neo-liberal agenda and urban spatial change signposted by cultural ensembles, designed by signature architects and enacted in global spectaculars. The globalizing project varies in detail by individual city but overall there are recurring features across the world including the reimagining of the city, the rewriting of the city for both internal and external audiences, the construction of new spaces and the hosting of new events. A major goal is the attraction of jobs and especially favored are the high tech and producer service sectors. Global city status is defined by having a range and density of symbolic analysts. A cosmopolitan lifestyle is also promoted as part of the project complete with settings and performance that synergize the four c's of culture, consumption, cool and cosmopolitan. The globalizing project also involves a spatial reorientation of the city, the spectacularization of settings, the creation of specifically global (in economic, cultural and political terms) sites and the encouragement of transnational locations.

The main thesis of the book is that this discourse of globalizing is a major force in the restructuring of cities around the world. The book will show how a similar range of globalizing practices – including the hosting of mega-events, the siting of urban spectaculars, the rewriting of the city and its representation to a world of global flows – are important processes structuring urban socio-spatial transformation in the contemporary world.

The book will draw upon a wide variety of cities in order to show both the similarities and the differences. Thus, while both Atlanta and Barcelona hosted the Summer Olympics with the associated reglobalizations of the respective cities, the redistributional consequences and spatial transformations were very different. Barcelona became a more public city while Atlanta became a more privatized city. The similarity/differences in the globalizing project is a useful way to compare and contrast the contemporary urban experience around the world.

THE STRUCTURE OF THE BOOK

Chapters 2 and 3 provide the theoretical formulations of the text. Chapter 2 examines the foundational works that have influenced the lines of recent research. This is not a hatchet job along the lines of "If only they were as clever as me." Rather I want to celebrate the early contribution of Peter Hall, John Friedmann and Saskia Sassen and show how they have influenced the course of what was known as world cities research. I look at the legacy of previous work and possible directions of future work. Chapter 3 picks up some of the possible themes that arise when moving away from traditional world cities research to looking at the globalization–urbanization connection through the lens of globalizing cities. Chapter 4 provides some theorized case studies of such an approach.

The next three chapters develop more theorized case studies of global cities. Chapter 5 flips the debate by looking at cities that are less globally connected, what I have termed the black holes and loose connections of the global urban hierarchy. It is as important to examine the less global as well as the hyper-global cities. The chapter is a useful corrective to the concentration on cities at the apex of the global city hierarchy. Chapters 6 and 7 look at the some of the most important characteristics of the more global cities. Chapter 6 identifies some of the main political-economic tensions of globalizing cities while Chapter 7 focuses on some of the key modalities of global cities.

Making the theoretical and empirical connections between the global and the local has been one of the more difficult areas of globalization studies and global cities research. In Chapter 8, I have attempted to explore some of the connections through an analysis of the Summer Olympic Games. Examining the Summer Olympics as a global phenomenon locationally tied in its performance to specific city sites allows us one entry into the nexus of city-global connections.

The remaining chapters are more speculative pieces, more in the way of introductions to possible avenues of research than fully finished conclusions to completed bodies of work. Chapter 9 raises in a speculative fashion, rather than in a detailed way, the role of the super-rich in shaping global city flows. This chapter will identify the global elites and especially the super-rich as an important motor force in the creation of global cities and especially in the creation of landscapes of conspicuous consumption within global cities. The final chapter raises the possibility of making connections between the global, the city and the body. Global cities research has scarcely considered the body while studies of the body often fail to consider the global and the city.

2 Globalization and the city

There are certain cities in which a quite disproportionate part of the world's most important business is conducted.

(Hall, 1966: 7)

The recent academic interest in global cities rests upon the work of pioneer scholars and researchers. Three important foundation studies were provided by Peter Hall, John Friedman and Saskia Sassen. Their work is important because they identified strands that, with one exception, can still be seen in recent research.

Peter Hall built upon the remark of Patrick Geddes made in 1915 that world cities were places were the world's business is conducted. In the first edition of *The World City*, Hall (1966) brought attention to cities as centers of political and economic power. He provided a detailed analysis of a selected group of what he termed world cities: London, Paris, Randstadt-Holland, Rhine-Ruhr, Moscow, New York and Tokyo. His focus was on the growth of these cities and the resultant planning issues and especially land use management. The history of ideas is full of instances where promising leads are forgotten. There was little discussion of why these cities were selected, but Hall's work was a genuine innovation. He emphasized the idea of the *global city as a planning problem* and focused on how to deal with population and economic growth while still maintaining a viable, livable city. Peter Hall also raised the issue of the global city as environmental issue. This strand was largely ignored in the subsequent world cities literature.

John Friedmann wrote two key articles. Friedmann and Wolff (1982) introduced the notion of a global network of cities as a research agenda for research. It was a political economy of the global nexus of capitalism and urbanism. The two authors connected patterns of urbanization to the internationalization of capital and urban restructuring to economic restructuring. They identified a global urban hierarchy with world cities at the apex. These cities were characterized as the control centers of the global economy, with a concentration of producer services, housing a highly mobile, transnational elite and the site of massive economic, social and physical restructuring.

World cities were the source of cultures and ideologies of consumption and global integration. Friedmann and Wolff identified the following as world cities: Tokyo, Los Angeles, San Francisco, New York, London, Paris, Randstadt, Frankfurt, Zurich, Cairo, Bangkok, Singapore, Hong Kong, Mexico City and São Paolo. These world cities were asserted rather than demonstrated. In a later paper Friedmann (1986) repeated the notion of world cities as basing points of the global economy and identified Tokyo, Los Angeles, Chicago, New York, London, Paris, Zurich, Rotterdam, Zurich and São Paolo as the first order centers of a global urban hierarchy. This was a similar but not exact list to that in the previous paper and just as based on simple assertion rather than careful documentation. Friedmann drew attention to the *global city as command center* and to the *global network of cities*. His loose definition of these cities would also be the hallmark of many subsequent studies.

In *The Global City* (1991) Saskia Sassen focused attention on the big three of London, New York and Tokyo. She examined their external connection to the global economy with emphasis on their command functions. She discussed global cities as centers for new systems of co-ordination and control, sites of production of specialized services especially producer services such as accountancy, financial services and consultancies. Another part of Sassen's work was to suggest that increasing social and spatial polarization was a feature of the global city. Even although the city was a shared space, it was a very differently experienced space. A powerful metaphor was the different uses of corporate offices; during the day housing the well-connected and well-paid making global transactions, by night cleaned by immigrant female workers paid minimum wages. Sassen drew attention to two strands, *the global city as command center* and *the global city as the polarized city*. Her emphasis on the big three of London, New York and Tokyo would also become a feature of subsequent research.

These three sets of foundation studies created a list of research topics that have persisted; four have retained their interest to researchers:

> The global city as planning issue
> The global city as command center
> The global network of cities
> The global city as polarized city.

In the rest of this chapter I will review some of the previous work in these four areas.

GLOBAL CITY AS PLANNING ISSUE

While the notion of the city as a planning issue has retained an interest, the notion of what constitutes planning has changed. Peter Hall was writing at a time (at least in the first edition of *World Cities*) and from a tradition that emphasized the active involvement of national and local governments to improve the livability of the city. Strands of Fabian socialism later morphing into a social democratic-liberal agenda were combined in Peter Hall's work with a belief in the role and efficacy of government and an assumption that the role of governments was to improve the welfare of its citizenry. This position has been undermined by a generation of Thatcherism, Reaganomics, the encroaching dominance of a neo-liberal paradigm and a jettisoning of the belief in the welfare function of the state in favor of the commitment to improving the competitiveness of cities. Planning has become a way to improve economic efficiency and market success rather than a process of improving social welfare and creating a fair and just city. Peter Hall's remarks now read like a voice from across a great divide, created by the tectonic shifting in political power away from labor to capital and from promoting welfare to improving market competitiveness. While the global city as a planning issue remains, the goals and techniques of planning have fundamentally shifted in two main ways: the political context of urban planning and the changing definition of urban planning.

There is no simple relationship between global cities and the state. In some cases, the city region may be separating off from the fortunes of the rest of the national space economy; in others the fortunes of nation-state and global city are inextricably linked. The different levels of the state are also involved in different ways. The relationship between the city and nation-state are many and varied. A post-Keynesian, post-federal state with marked inequality in the national space economy has replaced the Keynesian state committed to a large range of free public services nationally available. In states with a non-primate distribution and a federal structure, city-states can be identified in which the regional and city governments have become much more important as vehicles for globalizing cities. Take the case of Sydney, Australia whose Olympic bid was underwritten by the state government of New South Wales with the federal government playing little or no role. In other countries, those with a larger degree of central governmental power and a primate urban hierarchy, the national government along with regional and local governments can also play a part in ensuring the global competitiveness of global cities. The *grande ensembles* of Paris or the heavy UK government investment in London are two examples of national governments ensuring global city competitiveness.

We can increase our understanding of the global city by extending our understanding of planning beyond the manipulation of the physical environment to the conscious management of signs and symbols. An important element in global city research has been to identify the discursive strategies of global and globalizing cities (Ashworth and Voogd, 1990; Kearns and Philo, 1993; Ward, 1998). Global cities are represented by signs and symbols, advertisements and in the hosting of events. Repertoires of city advertising with their emphasis on economic advantages and quality of life factors have been identified (Short, 1999; Short and Kim, 1998). There are also cultural ensembles considered vital to global city status: art galleries, music venues, ethnic restaurants and festivals. Global cities are defined by cultural economic as much as financial economics. There are also constellations of urban spectaculars that include global mega-events and signature architects; Olympic Games, World Cup Soccer and music and arts festivals have become a defining feature of global city status. Global cities are enacted, performed and spectacularized.

GLOBAL CITIES AS COMMAND CENTERS

Global cities are significant locales and embodiments of economic and political power. In the late 1980s and early 1990s there was a plethora of studies that sought to identify global cities. Most of these studies were long on assertion and short on data. Data collection is a vital part of state surveillance and monitoring; so most good-quality data tends to be national while good-quality international data, such as that collected by World Bank, United Nations and the International Monetary Fund, tends not to be urban. At a very basic level it is very difficult and time-consuming to provide even a simple table of basic population totals for cities around the world that are comparable in terms of both time of collection and shared definition of urban, never mind anything more complicated in the way of sophisticated socio-economic data. There have been two responses to this data problem. The first has been to ignore it. This explains the large number of studies that simply present a list of global cities as given or make claims to global city status based on personal belief. Assertion rather than demonstration has been a major form of explanation in world city research: the so-called dirty little secret discussed by Short *et al.* (1996). The second response has been to identify and generate sources of comparative data. Short *et al.* (1996) and Short and Kim (1999) looked at three elements of economic power: stock exchanges, headquarters of major corporations and head offices of major banks. In contrast to most other studies this study examined data trends over time in order to get a feeling for the trajectory of change. While the position of New York, London and Tokyo was confirmed, the data also showed the relative decline of London and the relative growth of other European cities particularly Paris and Frankfurt. I would now theorize this finding along the following lines: we can imagine the world as a 24-hour clock. To get the full 24-hour coverage three world cities are necessary: one each in Europe, Asia and America. As the world turns, three cities strategically located can give global 24-hour coverage. When it is evening in Tokyo it is morning in London and when it is close of business in London, New York is already open for business and will stay that way until Tokyo picks up business again. The position of New York, the events of September 11, 2001 notwithstanding, is secured by the enormous dominance of the US economy. Tokyo's position is still relatively secure since Japan is the second-largest economy in the world, although the spectacular rise of China and Japanese deflation may be providing the scenario for more intense competition in the future between, on the one hand, Tokyo and on the other, Beijing and Shanghai. The Chinese competitive edge may be blunted by the present Chinese government's continuing commitment to Beijing while much of the commercial surge is centering on Shanghai. London, in contrast, is the capital of a relatively weak economy on the edge, in many ways, of an enlarging European Community. What the data was picking up was the competition for this central European world city slot. London is still dominant and, in terms of producer services, outranks most other cities in the world. But Paris and Frankfurt will compete in the longer term. The great silence of course is Berlin that for obvious reasons has not filled a peaceful global city role probably since 1936, the last Olympics in the city. Over the longer term, Berlin would be a strong candidate to become Europe's premier global city.

Subsequent analysis has led to two general conclusions: First, despite a slow dispersal over the past thirty years, a large proportion of command functions are still concentrated in just a few cities. Second, that rather than envisaging a model of competition between global cities for global dominance, it is more accurate to consider a network of global cities with economic control concentrated in only a few cities but with the operation of this power dispersed in the second and third tiers of the global city hierarchy through branch plants and back offices. New York, London and Tokyo continue to exist as centers of concentration of global economic command functions These three cities accounted for most of the trading activity in stock exchanges around the world. Together they provide 24-hour coverage, a continuous global coverage of market trading. They do not compete for business as much they complement each other, providing market opportunities for specific time niches.

While London's dominant stock market and global concentration of financial services is without parallel in the short to medium term, over the longer term, competition from European global cities such as Paris, Frankfurt and, more realistically over the very long term, Berlin, may force a change in the big three. Below the big three, national command centers in individual countries act as the main insertion point for global capital penetration. So while we have the global big three, we also have the national global centers such as Sydney which have few global command centers, in the form of head offices of global banks and corporations, but a large number of regional and local offices of global companies. The third tier of cities, such as Melbourne, to carry on with the Australian example, have smaller global command connectivities but still service regional and local markets. The global urban hierarchy is a network of flows that transmit global command functions across the world.

Studies of cities as economic command centers have identified a number of key characteristics. They have an articulated infrastructure of market trading that involves an agglomeration of demand and supply, an environment of innovation, business support services, a pool of highly skilled labor, a proximity of business organizations that allows information to be generated, analyzed and disseminated and deals to be struck. These are all set within a shared culture of expertise and a nexus of contacts set within an even broader context of a disciplined market, multi-currency trading, and a responsive central bank and governmental system of economic regulation.

A long time ago the English economist Alfred Marshall (1922) noted that there were three reasons behind the external economics of economic agglomerations: the pools of skilled labor that allowed the transfer of information, knowledge and skill; the presence of subsidiary industries that provided common goods and services; and the geographic proximity that facilitates face-to-face contact, the maintenance of trust and the exchange of information. Marshall's work has influenced much subsequent work and still provides the main theoretical understanding of the continued concentration of command functions in global cities. More recently, the concentration of economic command functions has been linked to the epistemic centrality of global cities. The globalizing economy creates lots of information, narrative uncertainty and economic risk that all have to be produced, managed, narrated, explained and acted upon. Global cities have become privileged sites of economic reflexivity, to use Storper's term (Storper, 1997).

Global cities are centers of global epistemic communities of surveillance, knowledge production and story-telling. Nigel Thrift has developed this theme in a discussion which, although widely pitched, is based almost entirely on London (Thrift, 1994, 2000). In an interesting although curiously placeless study, Thrift (2001) draws attention to the independent power of these cultural circuits of capital. He relates the boom and bust of the so-called new economy to the rhetorical fabrication of key actors, such as business schools, management consultants and management gurus. Epistemic communities not so much uncover reality as create it. Narration has consequences and consequences have consequences in the self-referencing world of financial epistemic communities.

A central reason for the concentration of command functions in selected global cities is the need for social relations in global financial business deals. Trust, contact networks and social relations play pivotal roles in the smooth functioning of global business. Spatial propinquity allows these relations to be easily maintained, lubricated and sustained. Global cities are the sites of dense networks of interpersonal contact and centers of the important business social capital trust vital to the successful operation of international finance.

One important strand in exploring the notion of global cities as command centers of the global economy is the discussion of advanced producer services such as banking, investment, consultancy, insurance and law. These services have globalized, or to be more precise, certain companies have offices in selected cities around the world. There have been a number of empirical studies of individual sectors of advanced producer services that highlight their global city bias. Beaverstock *et al.* (2000b) have looked at the geographical analysis of the 368 foreign offices US law firms. Just fifteen cities house 73 percent of the total, London alone accounts for 17 percent of all offices of US law firms abroad, and 59 percent are found in just eight cities: London, Hong Kong, Paris, Tokyo, Brussels, Moscow, Singapore and Frankfurt. The US law firms that had a global presence were concentrated in New York, Chicago, Washington, Philadelphia, Boston and Los Angeles. The same authors also looked at the location of London's law firms (Beaverstock *et al.*, 1999a). All but one of the top thirty London law firms had a foreign presence that in total had 221 foreign branches in sixty cities. Brussels was the most dominant city with twenty-five offices followed by Hong Kong (18), Paris (13), Singapore (12) and New York (10). The dominance of Brussels reflected the need for proximity to the EU Commission and its bureaucracy.

While the exact locational strategies may vary by individual sectors and national categories, it is clear that globalizing producer services firms tend to concentrate in a select range of global cities. The privileged sites of globalizing producer services are global cities. This tautology is deliberate. Globalizing producer firms create global cities; global cities attract globalizing producer services. Global service corporations have been adept at producing their own commodities, including new financial products, new advertising packages, new forms of multi-jurisdictional law. The one thing that all of these share is dependence upon specialized knowledge. Their state-of-the-art commodities are produced by bringing together different forms of expertise to meet the specific needs of clients. In order to be able to put together such packages, firms need to be in knowledge-rich environments. Sassen (1994) suggests that global cities provide such environments, and that face-to-face contacts between experts are facilitated by the clustering of knowledge-rich individuals in cities like New York, London and Paris. In this way, global cities have become "privileged sites" in the contemporary world economy housing the "knowledge elite" that enacts the economic reflexivity crucial for economic success. Reflexivity and networking are at the heart of understanding global cities as places through which people, institutions and "epistemic communities" work to establish and maintain contacts. More importantly, these communities act as crucial mediators and translators of the flows of knowledge, capital, people and goods that circulate in the world. A global city attends to the heterogeneous global space of flows, lending otherwise incommensurable materials intelligibility and translatability, e.g. credit rating mediates diverse banking systems while global law translates between different jurisdictions.

THE GLOBAL URBAN NETWORK

Ranking the command functions of cities is an interesting but limited exercise. It allows us some idea of the relative economic weight of cities and provides an antidote to the mere assertions of previous studies. However, cities not only occupy levels of a hierarchy they are also part of a network. Inter-city linkages take a variety of forms: economic, political and cultural. There are flows of goods and services, capital, ideas and trends, people and information that include foreign investments, overseas remittances, the tour paths of popular movies and pop groups as well as cultural flows including people, films, television, books and events.

Global urban networks have existed in the past. We have a marvelous visual record of one network, from a European perspective. By the last third of the sixteenth century there was a considerable stock of urban maps and images circulating in Europe. Compilations of city maps and prospects were published in 1551 and 1567, but the first city atlas was the *Civitates Orbis Terrarum* by Georg Braun and Frans Hogenberg. One volume was published in 1572 but it became so popular that by 1617, the work consisted of six volumes with over 363 urban views.

The first volume of the *Civitates* was published in Cologne, edited by Georg Braun and engraved by Frans Hogenberg; it contained prospects, bird's-eye views and plans of cities from all over the world. Braun (1541–1622) was a Cologne cleric who obtained maps and drawings from cartographers throughout Europe, commissioned work and wrote most of the descriptions. Hogenberg (1535–90) was an engraver. The atlas was meant to appeal to the lettered and unlettered. The Latin script appealed to the lettered but the views also made it accessible to the unlettered. Sometimes two sets of names were used for the one city – the Latin and the vernacular – so that the literate, but less scholarly, could see the name of their city.

Civitates drew from a variety of sources. Braun appealed to his readers to send him views and so the atlas grew with successive editions. The map of London, for example, shown in Braun and Hogenberg was based on the first surviving printed map done in 1553–59. It is referred to as the Copperplate Map. Completed at a scale of 25 inches to 1 mile it measures a grand 3.8 by 7.5 feet, covering fifteen separate plates. This map was the basis for the Braun and Hogenberg map of 1572, engraved by Hogenberg at a reduced scale of 6.5 inches to one mile.

Civitates provides us with a comprehensive collection of sixteenth-century urban views. Braun stipulated that the towns should be drawn so that the viewer could look into all the roads In the atlas each city has a brief written note of its history, situation and commerce. The prospect and the bird's-eye view predominate and even when the city is shown as a plan, buildings are shown in vertical relief.

In *Civitates* the city is both displayed and bounded. In almost all of the images, the city walls figure largely. Cities were often fiercely independent, the home to independent power centers, princes and prelates, guilds and town councils. Looking through the atlas at the many pictures of cities, one gains a very strong sense of cities standing apart, separate communities, reinforced by walls and battlements. The images also show the grandeur, wealth and power of the city. The city is not just represented but celebrated.

Many of the urban maps and views were made to evoke and represent civic pride. The atlas rejoices in the urban condition. Collectively, the images provide a comprehensive view of urban life in the Renaissance. They also indicate a world economy tied together in trade and linkages between urban centers. Aden, Peking, Cuzco, Goa, Mombasa and Tangiers as well as other cities around the world are represented. The global reach of mercantile capitalism and European colonization is evident in the range of cities represented in the *Civitates*. While the cities are depicted separately, the effect of the compilation is to reveal a global economy of urban nodes and a trading world of connected cities.

As a starting point for a discussion of the contemporary global urban network we can consider the work of Castells (1989, 2000) who describes contemporary society as a network society that operates in a "global space of flows." Castells argues that the spatial architecture of the world-system is based upon logic of flow, connectivity, networks and nodes. The informational global economy is one where core activities in the economy, in media communication, in science and technology and in strategic decision-making are linked worldwide in real time, creating the possibility of daily working on a planetary scale. However, Castells stresses that these decentralized networks of interaction require careful management to ensure their commensurability and functionality. The metropolitan nodes of global cities provide the crucial points of articulation in a society where there are possibilities of economic dispersal and distanciation. Two decades ago, the majority of corporate headquarters were located in global cities; now, new communications technologies do not require headquarter functions to be carried out only in the largest cities. In addition, the 1990s witnessed the rising importance of capital flows within so-called "emerging markets," bypassing the corporate headquarters of the urban West. There is now a space of flows where global cities are no longer defined by the presence of corporate headquarters.

We can build upon the Castells notion of space of flows, to examine specific flows in the contemporary global urban network. Again, data is a problem. Looking at international migration patterns, to take just one example, is bedeviled by lack of good-quality information. Some data sources have been constructed. Short and Kim (1999), for example, looked at airline passenger information to construct flow data based on passenger flows of more than 100,000 persons in either directions between pairs of cities. The overall trend was for the existing channel of flows between North America and Pacific Rim to thicken and deepen. It was not a global phenomenon. Like globalization itself the process was uneven, with Africa barely registering. Looking at the data for individual cities was revealing. London was clearly the hub of the global airline flows. However, comparison of secondary centers revealed a picture of regional articulations. Los Angeles, for example, despite the protestations of the LA School, is less a global city and more a Pacific Rim world city while Amsterdam is clearly a European world city. A most interesting pattern emerged from our analysis of Miami.

The data clearly revealed the city as the capital of Latin America. Miami is less a North American city and more a Latin American city with connections that make it the hub of flows to the Caribbean, Central and South America. Kingston, Jamaica, for example is divided into twenty postal districts. Locals use the term Kingston 21 to refer to Miami, indicating the degree of connection between the two cities. Miami is the global city of Latin America and the Caribbean.

Migration flows are the human face of globalization. Less than 3 percent of the world's population is a foreign migrant, but in absolute terms that constitutes a significant proportion, close to 180 million people. And this is probably a vast underestimate since the official counts fail to note illegal migrants. Global cities are the home to a variety of different nationalities. Transnational communities of both rich and poor of very different national and ethnic groups have been recognized in global cities. These communities are both the bearers and transmission lines of economic globalization. Much of the literature has equated economic globalization with the penetration of local markets by large multinational companies. This gigantist view of economic globalization ignores the extent to which economic globalization occurs in and through transnational communities in cities around the world. Transnational communities are the sites of linkage and flows between, on the one hand, the local and the family and, on the other, the global flows of people, money, capital, goods and services.

Global cities operate as major nodes of reflexivity in global networks through the migration of business elites. As the contemporary international service economy requires specialist professionals to be globally mobile to deliver intelligence, skills and knowledge to the point of demand, the development of a cross-border transnational migrant elite contributes to the production and consumption of the global city (see Beaverstock and Boardwell, 2000).

A discussion of network flows in the global urban hierarchy would be incomplete without mentioning the work of the Globalization and World Cities (GAWC) researchers who have constructed an invaluable data inventory (GAWC, 2002). To identify a world city network they looked at the distribution of advanced producer services across a range of cities. They generated a data matrix of 316 cities and 100 firms in accountancy, advertising, banking/insurance, law and management consultancy. They identified firms with at least fifteen identifiable separate offices. They identified connectivity between the 316 cities. Those that had at least one-fifth of the connectivity of the most connected city, which was London, were identified as world cities. A total of 123 world cities were identified (see Taylor *et al.*, 2001 who built upon the earlier 55-city network used in Beaverstock *et al.*, 1999b). They provide us with one of the most sophisticated world city networks produced to date. And even more reassuring, at least to this author, their data confirms some of the Short and Kim airline data that was constructed independently. Thus both the airline data and the GAWC data confirm the dominance of London, the relative provincialism of Los Angeles and the Latin American articulation of Miami's influence.

THE GLOBAL CITY AS POLARIZED CITY

In their foundational studies both Friedmann and Sassen drew attention to the social polarization in global cities. A compelling image is of the offices of successful financial services housed in global cities. By day they are full of high paid people on life's fast track. At night, low-wage workers, and often immigrants, clean them, with little job security and few benefits. The offices are like a metaphor for the city: same place but populated by people with very different life chances and experiences. Capitalism everywhere produces marked social inequalities. Is there something special about global cities that exacerbates social polarization? The deindustrialization in many global cities of the first world has reduced the number of relatively high-paying working-class jobs. Unionized jobs in factories provided steady incomes for many city residents. Their loss reduces the economic opportunities for many city residents. At the same time the increase in high-paying financial service jobs leads to groups of workers on the fast track of high incomes and generous benefits. These two trends tend to increase the disparity in economic opportunities for different groups in the cities. And even in the more lucrative financial service sectors, someone has to clean the offices. And office cleaners have rarely been as well paid as financial analysts or merchant bankers. Global cities can be sites of extreme polarization because the command functions pay well, the basic service sector pays poorly, while the deindustrialization in many First World cities reduces the job opportunities for many working-class groups. The disparities have been exacerbated in recent years because the shift from the Keynesian to the neo-liberal city has generally resulted in regressive social policies that reduce the social wage of the modest-income citizens. The neo-liberal agenda has dismantled much of the welfare state. The notion of dual city or divided city is an important theme of global cities research. Some differences have been noted. The strong welfare programs in many Western European cities compared to cities in the USA has meant a tighter compression of real incomes (wages plus the social wage). But even here the fiscal tax burdens of welfarism have generated some discontent amongst the business community. And the recent rise of an explicit anti-immigration racist political agenda in France as well as in such historically liberal societies as the Netherlands that focus in on generous welfare payments to undeserving others is an indication of the depth of popular discontent. Even if they fail to achieve political success, such parties and movements can play a significant role in shaping the public policy response of the political elites.

One argument is that urban politics in global cities may have a slightly different flavor because the sharp polarization in global cities provides for the emergence of new social movements that draw upon the marginalized and dispossessed (Keil, 1998; Keil *et al.*, 1996; Sassen, 2000). Purcell (2002) disputes this notion. His work on the rewriting of the city charter of LA finds little evidence of new social movements. New social movements may be more of a hope than an established fact.

THE SILENCE ON THE ENVIRONMENT

While I have reviewed the main articulations of global city research in the past twenty years, I want to end with one of the most obvious silences. The global city as environmental site has failed to attract the same attention as the other four topics listed on p. 10. The global city occupies physical space as well as discursive space; it is a physical entity as well as being a node of global connections. And yet the global city as an environmental issue is one of the great silences of the global city research activity of the past twenty years. Urban environmentalism is a neglected area in general, but there have been studies of urban environmental issues, especially of the large megacities in the poorer parts of the world with their horrendous problems of foul water and dirty air. But even in the richer countries the quality of the urban environment is an important issue of social concern and political mobilization. The lack of attention of global cities research on the urban environment is one of the most disturbing gaps in our understanding of global cities.

3 From world cities to globalizing cities

One important feature of world city research has been the search for world cityness. Many studies have been devoted to identifying whether this or that city is a world city. While the work is interesting up to a point it tends to focus on a narrow range of cities at the top end of the urban hierarchy. This focus tends to ignore how globalization is acting in and through all cities. There is a real need for extending the globalization/city research nexus beyond the restricted and constricting focus on determining which cities are world cities.

I want to shift attention from the measurement of world cityness to the examination of becoming a global city. Some of the contours of this change are mapped out in Table 3.1. By shifting attention away from the empirical measurement of degrees of "world cityness" I want to focus on asking what happens in all cities because of globalization. I use the term globalizing city to refer to the fact that almost all cities can act as a gateway for the transmission of economic, political and cultural globalization. The focus on globalizing city as opposed to world city shifts the attention away from the question of which cities dominate the global urban hierarchy to how all cities in the hierarchy are affected by globalization. In the rest of this chapter I want to explore some themes by identifying a list of possible research topics that could profitably repay some serious consideration. This list is neither complete nor exhaustive. Nor is it a prescription for other people's work. I want to encourage and enhance a plurality of voices and approaches. In the next chapter I will use this globalizing city perspective to provide a

Table 3.1 From world city to globalizing city research

World city research	Globalizing city research
Measuring globalization	Measuring and deconstructing globalization
Globalization	Globalizing
Measures of world "cityness"	Processes of globalization
City as impacted	City as arena
Being global	Becoming global

series of theorized case studies of Barcelona, Beijing, Havana, Prague, Seattle, Sioux Falls and Sydney.

CITIES AND CYCLES OF REGLOBALIZATION

Much of the globalization literature assumes that globalization is a relatively recent phenomenon. Elsewhere I have argued for the notion of waves of reglobalization since at least the late fifteenth century (Short, 2001). The past five hundred years has seen the growth of a functioning global economy. There have been a series of globalizations involving incorporation into imperial systems, attempts at economic decolonization and movement into and out of various global trading arrangements. While the latest round of economic globalization is particularly intense, marked by the creation of global markets, rapid capital movements, global shift in manufacturing, long and complex production chains and interlocking consumer marketization, it is only the most recent in a series of processes that have been occurring since 1500.

Globalization occurs in pulses, a series of reglobalizations that vary in form and intensity and leave an urban legacy of built form, social formations and new constellations of socio-economic power relations. Reglobalization is at times built into the form of the city itself. Over the centuries and through-out the many different political regimes, the layers of the city embody attempts by the controlling groups of the time to redefine or repackage the city according to the constraints and opportunities generated by the latest round of globalization. During this latest phase of globalization, for example, when tourist attractions are highly prized, many cities are repackaging the old with new accommodations or accessibilities to re-present themselves as living history and to take advantage of the global tourism economy. This new packaging reglobalizes cities in a distinctive way, building upon and partly erasing previous urban ensembles shaped by previous rounds of globalization.

Cities are not so much becoming globalized, as being continually reglobalized. The simple picture of globalization incorporating cities previously untouched by global connections needs to be replaced with a more sophisticated picture of economies and cities being subject to differing degrees and forms of reglobalization. An important research topic is to reconsider urban histories as part of a more global urban network continually in the process of reglobalization. The processes flow in either direction. One common model of globalization is as a wave of change sweeping away local distinctiveness. In this scenario, more often assumed than articulated, globalization is a tsunami of change wiping out the uniqueness of localities. However, a more critical view of globalization acknowledges a more complex set of relationships between the global and the local. The local is not simply a passive recipient of single, unitary global processes. Processes flow from the local to the global as much as from the global to the local (good examples are the growth in ethnic cuisines throughout the world and the blending of hybrid cuisines). The city is not simply a passive recipient of global processes. The term globalizing city refers to this more subtle relationship between the global and the local.

COMPETITION FOR GLOBAL CITY STATUS

Many countries have a stable urban hierarchy in which one city has dominated for years. However, in countries where there is not such a clear and overwhelming dominance there is opportunity for change and competition. The city that is most globally connected in any one country can change over time. For example, in Australia the primary global city has shifted over the past thirty years from Melbourne to Sydney. In Germany, Berlin is rapidly emerging as the primary global city. Examples of stable and changing primary cities can be noted and theorized. Taking the Australia case once again, the shift from Melbourne to Sydney embodies the shift from the importance of primary to tertiary goods and services in the Australian national economy's integration into the global economy. Melbourne was the headquarter center for national corporations involved in primary goods production, while Sydney has emerged as a major financial center. The shift from Melbourne to Sydney marks a distinct reorientation of the role and functioning of the Australian space economy in the global economy.

In China, there is a growing competition between Shanghai and Hong Kong for the global city designation. Hong Kong has long had an important connection with western economies long denied to Shanghai. However, that city is seeking to reclaim its pre-communist role as China's world business city. It has embarked on an ambitious rebuilding program involving high-rise towers in the financial district, deep-water container terminals, bridges and train systems. Hong Kong, in contrast, has been weakened by its physical and political distance from Beijing, soaring unemployment and plummeting property prices. However, Hong Kong still has a better legal system for fair business dealings and a higher quality of life. In 2001 the city introduced a new logo for the city, *Asia's World City*; alongside the words was a little dragon figure composed of two figures to represent East meets West. The branding of Hong Kong is part of the effort not only to gain greater international visibility but to win the national competition with Shanghai (*The Economist*, 2002).

REPRESENTING THE GLOBALIZING CITY

As cities compete to position themselves in the global flows of capital, images and narratives, selling the city has become an important part of urban promotion campaigns. Urban imagineering in the present era is dominated by selling the global connection (Short and Kim, 1998; Short, 1999). Place promotion involves the creation of a positive new image for cities to attract investment.

The specific promotional strategy of any one city depends on its product, or its existing resources. These include its culture and the potential for aspects of its culture to attract tourism as well as its locational resources such as low taxes, "right to work" laws and favorable geography for trade relations. The targeted audience of promotional efforts can be tourists, businesses or residents.

The reputation of a city, its image, is perhaps the most visible sign of promotional efforts. So important is the element of image to place promotion that Briavel Holcomb states, "the primary goal of the place marketer is to construct a new image of the place to replace either vague or negative images" (1993: 133). Images are presented to the world in TV ads geared toward potential tourists, in trade or industry magazines promoting business parks, or, increasingly, on websites intended for travelers, possible new residents or potential investors. It is not just global cities that are selling themselves in difficult times. Indeed, globalizing cities below the top echelon have a greater need to reposition themselves in the discursive space of urban imagery. The promotion of international competitiveness has come to be the hegemonic economic project for many cities around the world. A neo-liberalism now dominates the discourse of urban economic development and urban imagineering around the world (Hall and Hubbard, 1996; Harvey, 1989; Peck and Tickell, 1995; Short, 1999).

THE GLOBALIZING CITY AND GLOBAL SPECTACLE

Guy DeBord coined the phrase "society of the spectacle," asserting that "the spectacle is the chief product of present-day society," which is increasingly capitalist and global (DeBord, 1994: 16). The commodification of actual experience creates impersonal spectacles which are witnessed rather than experienced (Debray, 1995). Arguably, some of the most important global spectacles are sports mega-events such as the Olympic Games which reach a worldwide television audience and offer perhaps the best stage upon which a city can make the claim to global status. Presenting the host city with a unique opportunity to display itself to the world, such events, particularly the Olympics, provide an unsurpassed media spectacle focused on a distinct urban setting. The promise of worldwide exposure and economic gain has made hosting these major and regularly scheduled sporting affairs a lucrative goal for aspiring cities around the world

I will examine the global city connections of the Olympics Games in Chapter 8, but for the moment I will make some general remarks. The Olympics is a global media spectacle, a catalyst for urban change and a vessel for conveying and enhancing the host's cultural identity. The International Olympic Committee stated after the 1992 Games' spectacular ratings success that "it is through television that the world experiences the Olympics" (Tomlinson, 1996: 583). Yet the made-for-TV Olympic spectacle, epitomized by the opening ceremonies of each Olympiad, is a relatively recent phenomenon beginning in 1984 with the Los Angeles Summer Olympics. In attempting to represent the global Olympic ideal via the opening ceremonies, the host city presents its own version of that ideal, coating the global spectacle with the cultural flavor of the local host. Thus the Games' urban backdrop comes to the forefront of the spectacle's presentation and the host city serves as the nexus for the global and the local.

The Olympic city plays host to the world, "theatricalizing" the city and making it a media spectacle unto itself (Wilson, 1996: 603). As a vehicle for urban representation and landscape alteration, the Olympics and similar events contribute in various ways "to a profound shift in our relations to our urban spaces, spectacularizing them in the interests of global flows," often to the detriment of local communities (ibid.: 617).

As a factor in globalization, then, the Olympics and other global and regional media spectacles (e.g. the World Cup) have an immense impact on the urban image, form and networks of the host. They function simultaneously as an object of competition between cities, as catalysts for local change and as venues for establishing cultural identity through the "willful nostalgia" of a history created specifically for a global television audience (Maguire, 1994: 422).

THE CITY AND CULTURAL GLOBALIZATION

While the command functions of economic globalization have been extensively measured in the literature, rather less work has been done on measures of cultural globalization. The notion of the globalizing city provides us with a fertile ground for studies of cultural globalization, from empirically noting the changing level of "foreign" films in the city's cinemas and the number of McDonald's fast food outlets, to examining the role of cultural industries in the urban economy.

Theoretically, Arjun Appadurai (1996) proposes a way to conceptualize cultural globalization. He charts the global flows of culture as continually shaping and reshaping the world. He identifies five sites or realms in which these flows can be identified: ethnoscapes, technoscapes, financescapes, mediascapes and ideoscapes. These realms signify, respectively, the changes in the "landscapes of persons," technologies, finances or capital flows, the media, and the political configurations of such ideas as "freedom, welfare, rights, sovereignty, representation, and democracy." Together, they represent a tentative "model of global cultural flow" within which local practices and the movements of ideas can be positioned. Migration and media introduce and mobilize elements of culture, so that they circulate globally and become re-expressed through local contexts. Locality itself is a historical product. The processes that shape localities are not one-way interactions, but are rather dynamic and multifaceted, so that hybrids of the "newly arrived" and the "previously there" are constantly reconfigured and remobilized through global flows. Locality is produced through cultural practices.

Hybridity is a feature of globalized culture. For example, multinational corporations such as McDonald's, Guinness, and Coca-Cola are adopting practices of hybridization to make their global staples local favorites. Guinness hires locals in its factory in Accra, Ghana, McDonald's includes vegetarian menu items in India, and Coca-Cola features commercials from around the world, "set to the music of each locale" at its museum in Las Vegas (ibid.: 196). One of the guiding directives of presentation to these commercial giants is that "great brands are personal. They become an integral part of people's lives by forging emotional connections" (Rosenfeld, 2000: 193). Emotional connections are forged by engaging with local cultures and hybridizing global brands.

There is a connection between cultural and economic globalization. Behrman and Rondinelli (1992) argue that globalization puts pressure on cities to develop their specific cultures in ways that attract business, investment and high-tech professionals and that convince their own residents and entrepreneurs to remain.

RESCALING AND GLOBALIZATION

Much of the recent globalization literature has emphasized the apparent decline of the nation-state in the wake of increasingly fluid global capital flows. One common argument is that the state as it is presently formulated is simply unable to regulate or take advantage of globalizing trends (Ohmae, 1995). Such reports are greatly exaggerated. While it may be true that "electronic mass mediation and transnational mobilization have broken the monopoly of autonomous nation-states over the project of modernization" (Appadurai, 1996: 10), governments of nation-states may be far from ready to hand over control of their citizens to the flows of global influence. Though international pressures on such issues as human rights and democratic citizenship may be increasingly strong, states still have the necessary sovereignty to make laws regulating their citizens' behavior and civil participation. To varying degrees, according to place and time, states are still significant in the lives of their citizens. And, while some globalization processes may lessen this significance, some, such as the fears caused by large-scale immigration into a country, may also encourage a reactionary enforcement of state control.

Brenner focuses on the reterritorialization and rescaling of governance, identifying the global city as "the interface between multiple, overlapping spatial scales" (Brenner, 1998: 27). In asserting this, Brenner points out several flaws in the existing conception and discourse of globalization. First, this view neglects the "relatively fixed and immobile territorial organisation" of states that allows these processes to occur, and ignores the "major transformations of territorial organisation on multiple geographical scales" (Brenner, 1999: 432). The nation-state has not wilted in the sun of globalization. Rather, it has in many cases rescaled much of its authority to the local and regional level to take full advantage of globalization's benefits at the scales where the process is most active. The state has been "rearticulated and reterritorialized in relation to both sub- and supra-state scales" (Brenner, 1998: 3). This has translated into the relative gain of cities, especially those claiming global city status. Such a trend implies the "incipient denationalizing of select specialized national institutional orders" (Sassen, 1999: 167), particularly where global finance is concerned.

The European Union best illustrates the rescaling trend. The increasing integration of Europe's national markets has increased considerably the importance of Europe's major regional urban centers as competitive nodes in a growing urban hierarchy. The experience of a consolidating Europe displays the scalar dynamics and dialectic of territorial authority. The bulk of territorial authority is scaled down to the urban region level, a process allowed, facilitated and encapsulated in the ultimate spatial sovereignty of the national-state. Thus, "the success of local territorial competitive policies" (Cheshire, 1999: 861) becomes a key element in harnessing globalization. Competition between major functional urban regions, with cities clamoring to grab a bigger piece of the global economic pie and supported by state policy that increasingly makes them the primary actors, marks the current round of globalization in Europe as in much of the rest of the world.

A number of writers have drawn attention to the scalar dimension of understanding space and place (Brenner, 2001). Processes operate at a variety of scales: investment patterns, for example, simultaneously reconfigure the global economy, impact national economies and reshape regional production complexes. Specific cities, in turn, are both the setting and outcome of various scalar processes including global shifts in investment, national forms of regulation and local cultures of identity. We can theorize social processes as operating at various spatial scales, while we visualize space as embodying these multi-scalar processes and enacting inter-scalar connections. The processes and outcomes at particular scales elide and collide in complex patterns of cause and effect. Globalizing cities are sites of interaction between global processes and urban contexts as mediated through the state. Simplistic notions of the nation-state as overwhelmed by the forces of globalization, the so-called "hollowing out" of the nation, increasingly are superseded by more sophisticated models of the enduring yet protean state in a rapidly globalizing world (Swyngedouw, 2000).

THE CITY AND POLITICAL GLOBALIZATION

A global polity is a long way off. However, with the move away from the Keynesian to the entrepreneurial state, at both national and local levels, it is possible to discern new connections between political globalization and urban changes. One important change is the emergence of city-states separating from their national economies and the creation of city-to-city connections across international boundaries. There is almost a return to the era of city-states set in weak national systems of regulation as we witness the rise of urban economies separating from national economies as systems of national equalization are downplayed. Certain cities may emerge as almost separate economies from the national pattern and regional and urban–rural differentials may be exacerbated.

The transformation of finance, banking and business services, combined with the availability of new telecommunications technologies, has led not only to a concentration, but also to a massive decentralization that enables more and more cities and regions to become more connected to one another as they separate out from the less competitive and less globalized parts of their respective national space economies.

Increasing connections between cities across borders, such as sister-city projects, have led to the adoption of similar policies in urban management. Since 1950, 11,000 pairs of connections (sister cities) have occurred between 159 countries (Zelinsky, 1991). According to Zelinsky, sister cities have helped to foster a global village that ties in not only the movement of capital but also the growth of international tourism and sport; "Sister cities bereft of any historical ties can often ground their relationship on some shared social or economic interest"(ibid.: 21).

While much has been made of the disjuncture between cities and nations, global cities both reinforce and undermine this hypothesis. The fate of global cities are in some ways disconnected from their national territories, but in many cases the cities are so important to the national economies and national prestige. In primate city countries such as the UK and France, London and Paris represent national prestige and the urban face of the nation to the outside world. In more federal government structures with a less primate pattern, the fate of individual cities may rest with local and regional states rather than with central governments.

URBAN REGIMES IN GLOBALIZING CITIES

An urban regime is "the formal and informal arrangement by which public bodies and private interests function together to be able to make and carry out governing decisions" (Stone, 1989: 6). Urban regimes regulate the relationship between cities and the global economy. While much of the early work on regime theory was concerned with the internal dynamics of power sharing amongst disparate groups in the city, it is important to reconstruct urban regime theories to account for globalization effects.

At a fundamental level, in the current era of globalization, many of the governing decisions reflect an atmosphere of competition among cities and thus constitute an entrepreneurial model of regime (Elkin, 1987; Goodwin *et al.*, 1993; Harvey, 1989; Krätke and Schmoll, 1991; Stoker, 1990). An entrepreneurial city seeks to "facilitate privatization and the dismantling of collective services" in order to take advantage of the opportunities of connecting with the global economy (Lauria, 1997: 7). The role of political

governance in the city has been transformed by the rise of the entrepreneurial city. Urban regimes have become more concerned with direct
income generation and a variety of public–private partnerships. Cities have
become more concerned with the politics of maximizing growth and income
than with their redistribution (Hall and Hubbard, 1998). Developing urban
regime theories in studies of globalization, then, is to "note local and
national political differences that are capable of exerting significant
influence on the way globalization affects city development" (Leo, 1997: 78).
Some of the ways globalization affects development is, of course, social. As
a two-way regulatory body, or set of bodies, operating between state,
national and international structures on the one hand and local structures
or individuals on the other, an urban regime may have a direct role in
determining the extent of contact its citizens have in interacting with global
society. Lowndes recognizes that "while it is the nation state that ascribes
the status of citizenship to the individual, many of the rights and duties of
citizenship are exercised at the local level" (1995: 161). The extent to which
cultural globalization encourages a more even dispersion of democratic
values or to which it encourages controlling governments to further impose
regulations on the lives of its citizens has yet to be fully explored. A more
complex discernment of the role of urban regimes in regulating local structures will contribute to understanding how local and global forces interact
with each other.

GLOBALIZATION AND EVERYDAY LIFE

Globalization affects the institutions and structures of society, from multinational corporations to the range of opportunities and lifestyles available to
different individuals. Those living in global cities are certainly faced daily
with an increasingly "global" experience, but those living in smaller cities are
also finding differences in their everyday lives as globalizing processes occur.
Certainly, "globalization" is not sufficient to explain the many causal factors
that contribute to social changes, but the scales and spaces of everyday lives
are among the bundle of spheres touched by globalization.

As a set of processes, globalization triggers both new opportunities and
new problems experienced in local lives. Returning to an analysis of globalization in terms of changes in scales and spaces, we may ask how individuals
are at work in reconstructing the spaces of their lives and in turn, how the
spaces of their lives are being changed by globalization processes. For dis-
empowered or marginalized people, global technologies may allow for
interactions in more broadly defined or more diverse spaces. Globalization
may afford previously isolated small-town folk the opportunity to work as a
part of a large corporation and have access to increased wages and benefits.
Coversely, the closure of small, locally owned businesses and the inability of
family farms to survive may be linked with the incorporation of towns into
increasingly globalized markets.

Globalization also affects the degree to which local people or citizens have control over the identity of their places. The shifts in scales and spaces related to globalization are accompanied by shifts in power relations and in economic opportunities, and these may become manifest in individual lives in a rich variety of ways. World cities research as traditionally structured has tended to ignore the connection between global processes and local lives. The connections need to be made between the aggregate nature of most global cities research and the rich documentary evidence of urban ethnographies if we are to understand how different types of people get by in different ways in different cities across the world.

GLOBAL METRICS, NATIONAL DATA AND URBAN GROUND TRUTH

The connection between global processes, nation-states and cities and is made all the more difficult to ascertain because of the problems of data. At the national and international level this is has become less of a problem. There are now a series of global metrics. One of the consequences of political globalization is the construction and compilation of comparable national data sets. The International Monetary Fund, United Nations, World Bank and many other organizations such as the World Health Organization (WHO) provide global metrics based on national data collections. There are, of course, major issues of compatability and reliability, but, despite this, the data do provide a rough and ready method of intranational comparisons. The UN Human Development Report, for example, measures a country's achievements in terms of life expectancy, educational attainment and adjusted real income. High human development countries include Norway, Australia, Canada, Sweden, Belgium and the USA while some of the lowest are Burundi, Niger and Sierra Leone. The data have also been used in more diagnostic ways. The WHO, for example, has estimated that the minimum spending on health needed to provide "essential interventions" in poor countries is $40 per head: currently it is only $13. There are also more specific global metrics. The Civil Society Index, for example, plots the position of each country in relation to four dimensions of civil society: structure, environment, values and impact. Freedom House provides a measure of political rights and civil liberties that result in a crude tripartite categorization of countries into free, partly free and not free.

However, most of these data treat nation-states as single units, with one measure conveying the postion of the entire territory. As we have noted throughout this book, data is national rather than urban, yet people live in cities as well as being citizens of a nation. There are few sources of good-quality comparable urban data. One flawed exception is the Global Urban Indicators produced by the United Nations Human Settlement Programme. The program identified thirty urban indicators and nine qualitative data to be used in comparing cities across the world. The program provides a patchy coverage of these data sets for cities. So far the program is more of a promise of possible data rather than a delivery of global urban data.

We have better-quality national data than urban data and there is a mismatch between national and urban data even if both of them are of reasonable quality. To illustrate, let me use two data sets that measure globalization, one at the national and the other at the urban level.

The journal *Foreign Policy* publishes an annual globalization index (see www.foreignpolicy.com) that ranks countries. The Globalization and World Cities (GAWC) project ranks cities by their degree of global connectivity (the index is discussed in more detail in Chapter 5). While not exactly the same, they both measure forms of globalization. However, if we compare the rank of the country in the *Foreign Policy* 2002 index with the rank of the city in the GAWC index substantial differences can be noted. While Ireland was ranked first by *Foreign Policy*, the city of Dublin was ranked thirtieth by GAWC. The *Foreign Policy* index ranks sixty-two countries while the GAWC index measures 316 cities. To standardize, each index was divided into quintiles and only differences of more than one quintile were noticed. Thus Ireland and Dublin were in the first quintile of both countries and cities. More noticeable differences of three quintiles were recorded for South Africa and Mexico and two-quintile differences for Greece, Korea and Japan. In the case of the first two a low country score contrasted with a high globalization score for the city. In both these cases a low globalization index for the the nation as a whole data swamped the high index of the city. Both countries were in the fourth quintile for the country score, but in the the first quintile for the city score. A similar though less extreme pattern was noted for the other three where the country score was not matched by the global connectivity of the city. In other words, assuming a rough comparability between the data sets, aggregate national data sets masked the global connectivity of the cities within selected nations, especially those countries such as South Africa and Mexico with a very archipelago-type national space economy with sectors and spaces of high globalization set within sectors and spaces of relatively low globalization.

There is a real need to standardize our data and scale our theories so that we have both the appropriate data as well as the correct theory to uncover and explain the causal connections between global processes, national trends and urban impacts.

4 Globalizing cities

In this chapter I want to build on the ideas formulated in the previous chapter by presenting case studies of a range of globalizing cities. The cities and their respective population sizes are shown in Table 4.1. The number of the times each city was represented in fifteen "world city" studies as well as its composite index of world cityness, as devised by Beaverstock *et al.* (1999b) is shown in Table 4.2. Other cities could have been chosen but this is a representative sample of what we may term non-world cities as defined by the prevailing paradigm. Table 4.2 also gives the same data for London, New York and Tokyo to aid in comparison. The data shown in Table 4.2 reveal that the case study cities are not the usual suspects of "world city" research, yet, as I will show, they embody, reflect and transmit processes of globalization. Particular attention will be placed on processes of reglobalization in all the cities. I will also focus on spectacle (Barcelona, Sydney), economic centers (Seattle, Sioux Falls), urban regime theory (Beijing, Havana, Sioux Falls), cultural globalization (Prague) and urban representation (Sioux

Table 4.1 City populations

City	Earlier population (1987–1991)	Later population (1993–1998)
Barcelona	1,707,286[a]	1,505,581[a]
Beijing	5,970,000[b]	7,362,426[c]
Havana	2,125,000[b]	2,175,888[d]
Prague	1,214,772[e]	1,216,568[d]
Seattle	516,259[f]	536,978[f]
Sioux Falls	100,836[f]	116,762[f]
Sydney	3,097,956[d]	3,713,500[c]
London	6,378,600[b]	6,962,319[d]
New York	7,420,166[f]	7,322,564[f]
Tokyo	8,129,377[d]	8,021,943[d]

Sources: [a]Spain's National Institute of Statistics, 1990, 1998; [b]www.infoplease.com, 1987 Beijing and Havana, 1991 London; [c]UN Demographic Yearbook, 1997; [d]UN Statistics Division, 1989 Tokyo, 1991 Sydney, 1993 Havana, 1994 Prague, London and Tokyo; [e]www.encyclopedia.com, 1990; [f]US Census Bureau, 1990, 1998.

Table 4.2 Case studies cities and relative rankings of world cityness

City	Cited in literature (max = 15)	Index of world cityness (12 to 1)
Barcelona	2	4
Beijing	3	5
Havana	0	0
Prague	0	0
Seattle	2	2
Sioux Falls	0	0
Sydney	11	9
London	15	12
New York	15	12
Tokyo	15	12

Source: Beaverstock *et al.*, 1999b.

Falls). Each of the cities is discussed as both a unique place and as a representative of these themes. I have drawn on a wide range of studies to document the case studies; some of the more important sources are shown in Table 4.3.

SYDNEY

Sydney grew as an outpost of the British Empire, developed as a primate city, and more recently blossomed into a global city. An understanding of Sydney's growth and change allows us to see two things: how pulses of globalization affect one city over time; and the competition between cities for a country's global city designation.

Table 4.3 Case study city literature citations

City	Selected literature
Barcelona	Borja, 1996; Conversi, 1997; Hughes, 1992; McDonogh, 1987; Sanchez, 1992.
Beijing	Abramson, 1997; Dutton, 1999; Kaye, 1992; Leaf 1995; Pollock, 1997; Pomfret, 1999.
Havana	Coyula (Uggen, trans.), 1996; Halperin, 1994; Kaplowitz, 1998; Preeg, 1994; Preeg and Levine, 1993; Ritter and Kirk, 1995; Schwab, 1999; Segre *et al.*, 1997.
Prague	Gitter and Schueler, 1998; Simpson, 1999; Sykora, 1994.
Seattle	Friedmann, 1995; Gray *et al.*, 1996; Harvey, 1996.
Sioux Falls	*Christian Science Monitor*, 1992; Grimsley, 1999; Howlett, 1998; Merwin, 1983; Strout, 1999; Wilkinson, 1996.
Sydney	Birmingham, 1999; Connell, 2000; Moorhouse, 1999; Morris, 1992; Spearritt, 1999; Turnbull, 1999; Watson and Murphy, 1997.

The city first became global in 1788 when it was established as an antipodean gulag for the British state. The newly discovered country was initially a solution to Britain's overcrowded jails. Previously many convicts had been dumped in the North American colonies, but the US Declaration of Independence in 1776 closed off this possibility. The city's first permanent European settlers were convicts and gaolers sent to the other side of the world. Sydney was a carceral city, an outpost of the British state. Its connections were more overseas than national. In that sense it has always been a global city.

Through the nineteenth century its function broadened. It was an economic node in the British imperial system. It was a colonial entrepôt city, the transmission point between the wider world and the interior of Australia that was being commodified to produce wheat, timber, minerals and a range of primary commodities. These goods were sent to Britain, which in turn sent labor, capital and finished goods. Sydney was an important point in these economic transactions and the political and economic capital of the state of New South Wales.

Prior to 1901 Australia was in reality a collection of semi-autonomous states, including South Australia, Victoria, Queensland and West Australia. Each state acted as a separate economic and political unit with its own capital city; respectively, Adelaide, Melbourne, Brisbane and Perth. Though Sydney dominated the state of New South Wales, it was not the only gateway city in Australia. The other cities played similar roles for their respective states. Indeed throughout the latter half of the nineteenth century Melbourne could lay legitimate claim to being Australia's dominant city. After the Gold Rush of the 1850s in Victoria Melbourne's growth was spectacular and it was referred to as Marvellous Melbourne. When the new federal state needed a temporary capital, before Canberra was built, Melbourne was selected. In the period when Australia's globalization hinged around primary commodity production, Melbourne dominated. It housed the economic elite, and the headquarters of the commodity companies. Melbourne's claim to being Australia's global city was reinforced when it hosted the 1956 Olympic Games and 3,000 athletes came from sixty-seven countries.

In the second half of the twentieth century, the competition to be Australia's global gateway city was fought between Sydney and Melbourne. Each city represented the hopes and aspirations of their respective state governments, New South Wales and Victoria, as well as private sector interests. Each city had competing growth machines, to use Logan and Molotch's phrase. But in comparison to the USA, state governments played an enormous role in the growth machine and civic boosterism. Both groups realized that Australia could only sustain one global city. They both wanted it to be their city.

Sydney began to pull away from Melbourne in terms of international recognition. While both of them were approximately the same size (including the outer suburban areas, the figure is now approximately 4 million for

Sydney and 3.5 million for Melbourne), Sydney began to achieve more international visibility. The completion of the Opera House in 1973 gave the city a globally recognized icon. The project was begun in 1955 when the Danish architect Joern Utzon submitted the winning design entry. When it opened, the Opera House joined the Harbour Bridge in giving the city international recognition with a global signifier. For example, to celebrate the worldwide 2000 New Year celebrations the CNN cameras covered the Sydney celebrations. No other Australian city was visible to an international audience.

An economic shift also occurred when Britain joined the European Community in 1973. As part of the entry requirements, Britain had to jettison its old trading relations with Australia and New Zealand. Australia now had to operate in a global market rather than an imperial one, initiating a new round of globalization. In this new round of reglobalization, economic orientation shifted to the Pacific Rim and 60 percent of Australia's exports now go to Japan. There was also a reconnection with the world's financial system as Melbourne lost its national pre-eminence and Sydney emerged as Australia's premier, global gateway city. In the past twenty years Sydney has become the major destination for foreign investment and the leading choice for the headquarter siting of foreign banks, multinational corporations and high-tech companies. As Australia has reglobalized, Sydney has become the global gateway city.

The 2000 Olympic Games both embody and reinforce Sydney's position. In 2000 it hosted 10,650 athletes from 199 countries and gained global television coverage. The preparations for the Games, strongly supported by the State Government, included improvements to the international airport, transport connections between the airport to the city center and a host of infrastructural improvements. These and other remedial projects have further enhanced Sydney's global connectivity. The Games reinforced Sydney's dominance over Melbourne and accentuated Sydney's position as Australia's global gateway city.

BARCELONA

> The town had a gaunt untidy look, roads and buildings were in poor repair, the streets at night were dimly lit for fear of air-raids, the shops were mostly shabby and half-empty.
>
> (George Orwell)

George Orwell's 1938 portrait of Barcelona, torn and debilitated by the Spanish Civil War, contrasts sharply with the modern bustling image the city presented to the world at the 1992 Summer Olympics. At the opening ceremonies in late July, Mayor Pasgual Maragal declared to the world that Barcelona "is today *your* city" (Tomlinson, 1996: 597). The jump from the Orwellian world of the war and the Francoist state repression that followed to that of the Olympic Games is no small feat and illustrates the reglobalization of Barcelona and the city's attempts to earn "global city" status in the past quarter century. The themes of global spectacle and rescaling have marked Barcelona's recent trajectory.

Barcelona was the last Catalan stronghold of political resistance to fall in the Spanish Civil War (1933–9). Before the war, Barcelona had been the "*cap i casal* (head and home) of Catalonia" (Conversi, 1997: 35), a region sporting a long history of rivalry with Madrid and Castile. Catalan nationalism had thrived in the decades prior to the Civil War, but in 1939 Franco established a fascist state that emphasized Castilian control and Madrid's eminence in Spain. Franco moved quickly to quash Catalan nationalism, with particular emphasis on repression of the Catalan language.

The globalization of Barcelona since Franco's 1975 death has much to do with the unprecedented democratization of Spain, the resurgence of Catalan identity and the leadership of Barcelona's government. In two decades, Barcelona has changed from the dreary capital of repressed Catalonia to a shiny globalized metropolis. Franco's death ushered in a transition to democratization and decentralization in Spain that allowed regional identities to strengthen and flourish. The new national constitution gave Catalonia considerable autonomy. Barcelona again openly celebrated and dominated Catalan identity as new political actors took the stage in the region and began to push Barcelona toward its globalizing goal. Accompanying this assertion of Catalan identity was a further decentralization and democratization of government on the municipal level within Barcelona. The devolution of government to the *barrio* (neighborhood) in Barcelona and its metropolitan area encouraged local initiative, the greening of the city and a strong sense of civic pride and unity, resulting, some have argued, in economic revival, improved public services and metropolitan coordination on large projects (Borja, 1996: 85).

Perhaps most important to Barcelona's globalization was Spain's 1992 entry into the European Union. Catalonia, with six million people, 21 percent of Spain's GNP, over a quarter of its exports and a third of its foreign investment, has firmly established itself as a vital component of both Spain and the new integrated Europe (Thomas, 1990: A1). Barcelona, as the gateway to a prospering Catalonia, is reaping the rewards of integration and repositioning itself within Spain, the Mediterranean and Europe.

The 1992 Olympics set a number of records, among them the most television viewers at 3.5 billion worldwide. The Olympics offered Barcelona the opportunity to make great changes in its physical appearance as city leaders undertook numerous large urban projects: Montjuic Stadium, the Olympic village, and the Contemporary Art Museum were accompanied by many smaller urban projects designed to "green" the city and make it more pedestrian-friendly. The dramatic architectural and planning changes undertaken in Barcelona earned the city the 1999 gold medal from the Royal Institute of British Architects, the first time that award had been given to a city rather than to an architect.

Urban iconography is a powerful component of city representation, and in Barcelona's case the works of Catalan architect Antoni Gaudí epitomized the city's unique culture and distinctive built form. Working in the late nineteenth and early twentieth centuries, Gaudí was a leader in the architectural style *modernisme*, distinctively Catalan and marked by fluid shape, bright color and a highly original feel. Gaudí's architecture is the touristic platform of Barcelona's urban form. The 1992 Olympics provided an ideal arena in which to celebrate, update and expand this tradition.

The Olympic spectacle and the physical changes accompanying it remain building blocks for Barcelona and the high point of its process of reglobalization. Barcelona's role at multiple scales (Catalonia, Spain and Europe) retains great importance, though Barcelona has yet to reach the world status of other European cities such as London and Paris.

PRAGUE

The cities of post-Soviet Europe are important sites of reglobalization into a capitalist global economy. As just one example of this process we will focus on Prague and its attempt to build on its rich cultural heritage to become a global tourist center.

Prague's built form is a museum of almost a millennium of European architectural styles including Romanesque, Gothic, Renaissance, Baroque, nineteenth-century revivals of all of them, and Art Nouveau. Alone of large central European cities Prague escaped unscathed from the devastating city bombing campaigns of the Second World War. Some of the most notable architectural ensembles include the Prague Castle, Charles Bridge, Old Town Square and the Astronomical Clock, Wenceslas Square and the Old Jewish Quarter. The city has long played a key role in the European cultural scene. The eighteenth century brought about an era of classical music in Prague which saw several new municipal opera houses including the F.A. Nostitz theater (later renamed the Stavovske theater). It was here that Mozart premiered *Don Giovanni* (1787) and *La Clemenza di Tito* (1791). The city is also well known as the home and burial ground of Franz Kafka.

Some 215 years after *Don Giovanni* premiered in Prague the city has once again been recognized by its European neighbors as a cultural mecca. The European Union Council of Ministers of Culture selected Prague as the European City of Culture in 2000 (European City of Culture, 2000). This prestigious award is given annually to the city that best exemplifies European cultural heritage, the first being Athens in 1985. The honor of being named the European City of Culture had great significance for the city in terms of tourist publicity and as an active element in its shedding its more isolated communist past. The city has sought to position itself in public awareness away from a place identified as formerly East European towards becoming the new capital of central Europe.

Tourism has become an important element in the city's capitalist revival and reglobalization. The increase in tourism has had positive and negative implications for the character of the region. Even though residents welcome tourist activity they shun the increase of tourist retail shops in the city center; nearly half of residents feel that there were too many souvenir shops in the historic core (Simpson, 1999). To increase the ambience in the area, a reduction in the number of tourist users was more strongly supported by visitors (77.2 percent) than by residents (34.3 percent). Tourists are beginning to feel that they are not seeing the "real city" or mingling with the locals, features which are advertised as being Prague's charm.

Prague is now at a crossroads in its reglobalization. On the one hand, the city is experiencing tremendous economic growth through its expanding tourist industry. On the other hand, the influx of tourists is threatening to destroy the very city they have come to see.

SIOUX FALLS

Globalization is truly a global phenomenon that impacts more than just the largest cities. Though small and rarely appearing in tables of "world cities," Sioux Falls is also a crucible of globalization trends. Sioux Falls has long been the largest city in South Dakota. Labeled "The Gateway to the Plains" by the Sioux Falls Development Foundation and city marketers, Sioux Falls is an example of how the idea of a globalizing city can be applied to a small-scale regional city. In recent years, Sioux Falls, promoted by a vigorous boosterism campaign, has been making headlines and ratings. Sioux Falls has received accolades from a range of sources as widely read as the *USA Today* and as specific as *Builder Magazine* (see Table 4.4). Behind these acclaims and statistics is a narrative that has been managed by city leaders to take advantage of globalization.

Table 4.4 Sioux Falls recognitions

Designation	Source
Best place to live in America	*Money Magazine*, 1992
No. 1 economy for 4th consecutive year	*Money Magazine*, 1994
Nation's 7th-best city in which to raise a child	*Parenting Magazine*, 1997
One of the top 10 cities with the friendliest environment for working women	*Redbook*, 1996
4th-hottest city for selling	*Sales & Marketing Management*, 1999
Lincoln County – 2nd fastest growing US county	*USA Today*, 1998
One of the next big growth markets for residential building	*Builder Magazine*, 1998
One of 3 best small cities in which to start a business	Cognetics, Inc., 1999

Urban regimes and strategies of place promotion are among the key tactics cities use to connect with global capital. In Sioux Falls, alliances between the private business community and the local and state government are specifically designed to attract new businesses and residents and to make Sioux Falls more competitive. The Sioux Falls Development Foundation, which describes itself as "a non-profit development corporation that facilitates the attraction of new businesses, the retention and expansion of existing firms, and the formation of new companies" (Sioux Falls Development Foundation, 1999: 1), is a key actor in the globalization of Sioux Falls.

One of the more common activities of cities in a "footloose" globalized economy is to enlist the services of a location consultant. The Sioux Falls Development Foundation has retained the services of a location consultant, the Boyd Company, to conduct studies of the relative benefits of doing business in Sioux Falls as compared to American and Canadian cities in the Midwest region. The reports of the Boyd Company are reproduced in such trade-specific publications as *Modern Plastics*, which announced in 1993 that, "the site where operating costs are the lowest for plastic processing plants is Sioux Falls, South Dakota" (Toensmeier, 1993: 1). The spread of such business-friendly reports is one of the primary aims of the Development Foundation. According to the Foundation's marketing director, "Ours is an on-going effort in order to educate the corporate world that Sioux Falls is a great place to do business" (Hindbjorgen, 1999). The strategies of the Foundation are clearly paying off, since *Site Selection Magazine* named the Foundation-owned Sioux Empire Development Park one of the top ten industrial parks in the world.

Despite the boosterism, there are segments of the population not benefiting from the growth of the city. In 1998, 432 homeless people were counted in Sioux Falls, 82 more than the previous year. In fact, "the number of homeless in Sioux Falls jumped dramatically after *Money Magazine* named the city the best place to live in America in 1992. The opportunities in Sioux Falls are not benefiting everyone. A series of articles in the *Argus Leader* described the conflict over the use of downtown Sioux Falls, which has been altered significantly by the opening of the Washington Pavilion of Arts and Science in 1999. Making the cultural consumers who visit the Pavilion comfortable, though, means regulating the activities of youth who have a tradition of making use of downtown Sioux Falls by "cruising The Loop." Ideas suggested by members of the Downtown Task Force, composed of government officials, downtown merchants and residents, include measures to ban cruising and the installation of surveillance cameras on downtown buildings. The changes in the streets of Sioux Falls signal tensions between the small-town practices that once defined the area and the ambitions of business and community leaders who want to redefine Sioux Falls as an urban cultural center for a selective audience.

Sioux Falls' leaders have sought to further define the city as a viable part of the global economy by establishing a US port of entry and a foreign trade zone. The Governor of South Dakota, Bill Janklow, lobbied for these features in Sioux Falls to encourage local business owners to have international deliveries made directly to Sioux Falls rather than going through Minneapolis or one of the other larger regional cities. In local newspapers, Janklow expressed hopes that the foreign trade zone would convey the message that Sioux Falls is a legitimate international market and a gateway city to the global economy. In agreement with Janklow, South Dakota Senator Tom Daschle said of the opening of the Port of Entry: "This development is a great start for the new year for South Dakota businesses competing in the global marketplace, and I hope more state businesses will resolve to get involved in international trade" ("Sioux Falls Designated . . .," 1995).

By taking advantage of their location in a low-tax state and through an aggressive place-marketing campaign, the leaders of Sioux Falls have successfully established the city as a gateway to the global economy in the Great Plains region. The city is connected to the global economy through business negotiations and through transactions at the new port of entry and foreign trade zone. For individuals in Sioux Falls, there are new spaces, such as the revitalized cultural downtown, and new scales, such as overcrowded homeless shelters and large corporations, that are becoming part of daily experiences. The example of Sioux Falls shows that globalization impacts small cities, as well as large cities.

SEATTLE

The World Trade Organization (WTO) meetings that took place in Seattle in November 1999 became a scene of massive protest against what critics claimed was an unfettered global capitalism. What was meant to be a quiet economic trade meeting became a boisterous global spectacle of resistance to globalization. The city embodied the strains of global capitalism.

The WTO meetings were not the first time that the city has hosted a trade conference. During the 1990s city leaders actively sought trade conferences and global events to help highlight the city as a global arena of commerce and culture. Conferences have included the Quadrilateral Trade Ministerial meeting in 1996, the Asia Pacific Economic Cooperation (APEC) summit in 1993, the first US meeting of the North American Free Trade Agreement (NAFTA) in 1990 and in 2001 the Asian Development Bank Annual Meeting.

The city also has an active trade alliance committee (Trade Development Alliance of Greater Seattle) that forms ties with other cities in the region and abroad. The "Cascadia Region" that Seattle has formed ties with is a 400-mile long, 8-million resident corridor that runs from Vancouver, BC to Eugene, Oregon. It accounts for more than $250 billion in annual output; if ranked as a nation the region would be the tenth largest economy in the world (*Economy*, 2000; Harvey, 1996). The governments of the region "have agreed to cooperate in areas such as transportation planning, trade and tourism promotion, border crossing improvements, economic development, natural resource management and special events" (*Economy*, 2000). In conjunction with regional alliances the city has been an active proponent of international alliances. Over the last forty years an extensive sister-city campaign has been conducted that now includes twenty-one sister cities, making it the second-largest such effort in the USA. The sister-city program helps to establish not only economic but also cultural links between cities.

Companies such as Amazon, Boeing, Weyehaeuser, Microsoft, Nordstrom and Starbucks all have their headquarters in the Greater Seattle region. The world's software giant Microsoft is just one of 2,200 software development firms in the area. From the local to the global: Starbucks was once a small coffee shop that opened its doors in 1971 in Pike Place Market. The company has prospered along with the city of Seattle, and is now a global empire that has approximately 2,200 stores worldwide.

While Seattle's exploding global economy has brought many economic and cultural benefits to the area it has also come with a downside. As more and more people have moved into the area, rising housing prices and rates of congestion have become a major concern. Affordable city living has become a thing of the past. For instance from 1988 to 1998 the median sales price of existing single-family home in metropolitan Seattle area increased by 86 percent, compared with the national average increase of 46 percent (Caggiano, 1999).

Along with a massive increase in the housing prices the area has experienced increased congestion. The area's freeways during rush hours have become a bottleneck of cars. This has meant a nasty commute for many who, due to metropolitan housing prices, must live outside the metropolitan region and commute into the city for work. In a 1997 ranking by *The Economist,* Seattle was rated as having the sixth-worst traffic in the country.

The boomtown effect is not covering the entire Puget Sound region. While Seattle is booming, the city of Bremerton, just a short ferry ride across the Puget Sound, is on the decline. Bremerton is a traditional "blue collar" machinist/navy town, that has experienced the downside of Seattle's boom. The same things that have helped Seattle to foster growth such as trade liberalization and the ending of the Cold War have had a negative impact on Bremerton through shop/base closings and a loss of well-paying "blue collar" jobs.

As with globalization everywhere, the effects have been uneven in spatial and social terms. However, the dominant narrative is of the Seattle of Starbucks and Microsoft rather than of Bremerton and low-income households squeezed by expensive housing.

BEIJING

Today's Beijing is awash with change, where the old Confucian ideals of personal cultivation and family values clash with a new emphasis on money and the market, where a bureaucratic culture designed to hold people in check is replaced by a rootless mobility, where a construction boom is reshaping Beijing's low-slung profile and cramped alleys with soaring skyscrapers of glass and steel, where car traffic clogs streets that once rang with bicycle bells, where dust mixes with vehicle exhaust to form a near-constant polluted haze, where unemployment and gross underemployment create unease, where corruption has brought thousands of protesting citizens into the streets.

(Carrel, 2000: 120)

Tension and contrast mark the reglobalizations of cities. The latest phase of Beijing's reglobalization is no exception. According to journalist Todd Carrel, "After watching China's economy struggle for decades, Communist Party leaders took a gamble: Could they relax economic controls yet still manage to keep a firm grip on political power?" (ibid.: 117). The city's re-globalization is evident in the built form of the city and the successful government-led committee trying to secure the 2008 Summer Olympic Games. The government-sponsored website for Beijing lists building projects open to investors and the Beijing Urban Development Corporation claims it is "ready for cooperation with the parties from other parts of China and abroad to make greater contributions to the urban construction of the Capital of China" (www.beijing.gov.cn). This openness to foreign money is in conflict, though, with a Chinese government that still wants to control its people and maintain a distinct Chinese identity. Foreign investment, largely from ethnic Chinese living abroad, has financed an increasing number of development projects, and in turn the Chinese government seeks to improve the environment in order to continue to attract foreign investment. The opening of the market and the influx of foreign money has led to higher circulation of money and increased housing prices. Leaf's (1995) study showed that the increased marketization of housing was leading to greater spatial segregation by income class in Beijing. The city government is relocating approximately 2.5 million of its 11 million residents to suburbs, in part to "make room for new tourism centers, expensive apartment com-pounds, and department stores where Beijing's new wealth and foreign investment dollars are getting spent" (Carrel, 2000: 124). The opening of China to the global market and the resulting influences of foreign money and interests are changing the state-enforced spatial equality that China was ideologically committed to during the peak of Maoist Communist rule.

During its unsuccessful bid for the 2000 Summer Olympics, Beijing officials exercised strict regulation of the city's appearance in an effort to impress a positive image upon the International Olympic Committee. Beijing was rejected as a host city for the Olympics because of objections to civil rights violations in China. Though its slogan in the Olympic bidding was "A More Open China Welcomes the 2000 Olympics," evidence suggests that policies of strict social regulation shrouded even the initial inspection visits in the city (Kristof and Wudunn, 1995).

The opening of China to the outside world has led to tensions between the old city and the new, between the government and the international community, between the government and citizens, and between generations. Beijinger Wang Yingchuan, when asked what the youth of the city wanted, replied, "to get on the Internet, to play sports, to dance at the discos" (Carrel, 2000: 137). Beyond Beijing's youth, global culture has reached the new success stories of the global economy. In downtown Beijing, "entertaining well-to-do investors, party bureaucrats, and a budding cadre of entrepreneurs requires something more than operas preaching the glories of socialism. At a dinner theater modeled after the Moulin Rouge in Paris, showgirls perform French-style cabaret, complete with the cancan"(ibid.: 120).

The government's recent treatment of members of the banned Falun Gong spiritual movement suggests that the city's accommodation to diversity does not go so far as to include what is seen as a threat to the unity of the nation or the power of the Communist Party. China outlawed the group in July 1999 and officials have since been detaining Falun Gong members.

Used during the Ming and Qing Dynasties for grand ceremonies and by Mao Zedong in 1949 to proclaim the founding of the People's Republic of China, Tiananmen Square in Beijing has frequently been a symbolic center of Chinese political power. It has been the site and setting of numerous debates about the meaning of China. Tiananmen Square has also been the site of contestation, most notably in the case of the 1989 pro-democracy student protests and more recently in the case of Falun Gong supporters. The Square is still promoted, though, as a center for tourism and unified Chinese culture, and "no monuments in Tiananmen Square mark the massacre" of 1989 (ibid.: 137). For the fiftieth anniversary celebration of the People's Republic in Beijing, government officials removed the scars of tank treads from the 1989 massacre from the road in front of the Gate of Heavenly Peace. Reglobalizations often involve erasures as well as inscriptions. Beijing has been a world city for centuries, and today it is going through another phase of reglobalization. Adjusting its isolationist policy to accommodate the opportunities of a global economy, the Chinese government has been trying to control the cultural and political consequences of globalization. As the Chinese government works out its relationships with the international community, Chinese people will be working out their relationships with their government, and Beijing will continue to be a focal point for the manifestation of the tensions of globalization.

HAVANA

On January 25, 1998, Pope John Paul II delivered a lengthy open-air Mass to hundreds of thousands of Cubans in Havana's Plaza of the Revolution in which he encouraged Cuba to "open itself up to the world" while inviting the world to do the same for Cuba (Schwab, 1999: 126). Havana's experience in the last decade illustrates its hazardous road toward reglobalization and the dangers of being left out of the world economy.

Havana has been a global city since the Spanish founded it in 1519. By the early seventeenth century Havana had become the primary city in the Spanish Caribbean, serving as the commercial and military pivot of Cuba and the principal node in the trading network connecting Spain with its New World colonies (Segre *et al.*, 1997: 19). The sugar boom of the nineteenth century helped finance extensive urban projects in Havana, particularly during the 1830s under the direction of colonial Governor Miguel Tacón, and by the 1860s urban sprawl forced leaders to tear down the city walls (ibid.: 33). Havana expanded rapidly after the Spanish–American War in 1898. With the influx of American tourists and investment, a definite Yankee influence emerged in the newer suburban quarters of the city. Firmly within the American orbit by 1898, Cuba and especially Havana were dominated by American investment until the late 1950s. Havana was a secure investment for US capital, a playground for American tourists and a haven for organized crime.

When the Cuban Revolution finally succeeded in January 1959, Havana's global connections altered suddenly and drastically. The Marxist-Leninist state headed by Fidel Castro established ties to the Soviet Union and its European satellites. Official links to the United States ended as the American embargo against the island took hold. Cuba was financially kept afloat by the USSR that bought up all of the national sugar crop. The "Special Period" that followed the end of the Soviet connection brought to a head the urban problems that had been building in Havana for decades. In the past decade Havana has faced many of the same problems confronting other large cities in the Third World. Yet Havana is unique because of the persistence of Castro's socialist government and the state's open hostility toward the primary channels and attributes of economic globalization. At a January 2000 conference in Havana, Castro labeled the IMF "the backbone of the New World Order of Globalization" and the "executioner" of the Third World (Anderson, 2000: 224). Cuba's resistance to traditional Third World links to the world economy meant alternative routes to reglobalization and reintegration had to be found. The desperate search for new global links made the period of economic restructuring in the early and mid-1990s a desperately hard time for Havana.

By the end of the decade, these links had been established in Havana through the growth of tourism and an emphasis on neighborhood improvement. Many analysts predicted that the only solution for Cuba's economic woes would be a lifting of the American embargo. There are signs that the embargo, while still in effect, has weakened. The primary goal of the embargo, the ousting of Fidel Castro, has been unsuccessful. Increased American pressure and attempts to expand the embargo failed as well in the 1990s and only served to alienate American allies, the UN and other international groups (notably the WTO and Organization of American States) from Washington's position. Furthermore, American economic penetration has not been stopped within Cuba as dollars have flowed into Havana via the tourist industry and Cuban-Americans sending money to relatives at home. In fact, US dollars have been accepted in Cuba as legal currency since 1993.

Marginalized by the USA, Cuba has still managed to maintain connections with the global community. The United Nations has taken a further interest in historic Havana, naming Old Havana, the city's colonial core, a UNESCO World Heritage Site in 1982 (Williams, 1999: 38). This created a widespread interest in the preservation and rehabilitation of Havana's historic districts.

The Pope's 1998 visit to Cuba marked a watershed in the reglobalization of Havana. The pontiff's appeal for the opening of Cuba to the world, and the world to Cuba, symbolized the tensions in Cuba's and Havana's new round of reglobalization after the breakdown of the Soviet alliance.

CONCLUSIONS

This chapter has argued that to fully understand the connection between globalization and the city, it is important to extend our understanding beyond the narrow focus on the usual suspects of large global cities. While the top level of the global urban hierarchy is an important object of consideration, when it becomes the sole focus of understanding the globalization–city nexus then understanding is skewed and partial. Theories of globalization that only build upon the experiences of a few global cities have a precariously narrow grip on the full range of the urban experience, while the arid search for world cityness dooms a large number of cities to marginality or even exclusion from research on globalization and the city.

Shifting attention away from identifying a narrow range of world cities to a more inclusive concern with globalizing cities extends the range of theorizing on how globalization takes place. Globalizing city is a shorthand term for the idea that many, if not all, cities act as transmission points for globalization and are the focal point for a whole nexus of globalization/localization relationships. We selected seven cities ranging in population size from just over 100,000 to almost seven and half million, and ranging in "world-cityness" measures from relatively high to not even registering. We purposely selected cities that were not on the usual list of world cities and below the top echelon of the global urban hierarchy. Our case studies could have been different but the general point remains that even small, so called "non-world" cities can be examined for evidence of globalization. The case studies were brief. Each city could have been the focus of the entire book. However, they were indicative of the rich possibilities of using the globalizing city themes and the selected topics. Globalization is a phenomenon that is occurring around the world in a range of cities. By focusing on the idea and practice of globalizing cities, our understanding of both globalization and the city can only be enriched and deepened. The theme of globalizing cities lays the basis for a sounder theoretical understanding of the impact of globalization on different cities in the world and for a more profound explanation of the connection between urbanization and globalization.

5 Black holes and loose connections

Globalization is an uneven process. Cities are reglobalized in different ways at varying rates. While much of the recent literature has focused on measuring the converging points of globalization, I want to change the angle of vision and focus instead on identifying cities that seem to be bypassed by at least some of the more advanced forms of economic globalization.

Following on from the early work of Hall (1966, 1984) and Sassen (1991), there has been substantial work on the notion of global cities. They have been theorized as the command centers of the global economy, vital hubs in the flows of goods, people and ideas. However, there are three problem areas in this research area. The first is that work is limited by what Short *et al.* (1996) refer to as the dirty little secret of world cities research, which is the lack of good-quality, comparable, international urban data. The data deficiencies inhibit sound theorizing. Second, the search for world cityness in a range of cities has bent the research work towards the top end of the urban hierarchy and often limited the discussion of the connections between cities and globalization to a search for only a narrow range of world city functions. Our understanding of the connections between globalization and the city has become biased towards only looking for the urban impacts of globalization in the big world cities. Because a city is not designated as a world city does not imply a lack of global connections. The designation, in effect, refers only to the command and control functions of the global economy. A city not designated as a world city can still be actively involved in global flows of goods and services and still serve as a hub in the flow of people, remittances, finance and ideas. Third, the research has concentrated on searching for evidence of global connectivity. But, while it is important to identify global cities, it is also important to identify the lack of connectivity. Silences are as interesting as utterances. They tell us much about the process of globalization and development. In this chapter, preliminary data analysis is presented to identify black holes and loose connections of one global urban hierarchy.

BLACK HOLES

The very largest cities in the world are also some of the most globalized. However, not all large cities are global cities (Taylor, 1999). To identify very large, less globalized cities two data sets were combined. The first is the population figures for major agglomerations made available in Brinkhoff (2001), who provides the most accessible and up-to-date population figures for urban agglomerations around the world ranked by population size. The largest is Tokyo with a population of 34 million, followed by New York at 21.5 million and Seoul at 20.4 million. Population figures for metropolitan areas are notoriously suspect. Identifying the functioning urban region, as opposed to the formal jurisdiction, is a difficult matter. Moreover, national differences in the definition of urban make it difficult to compare population figures for cities in different countries. The fact that the "Brinkhoff" population figures are all round numbers, the figures for Los Angeles, USA and Mumbai, India, for example, are rounded to 16,700,000 and 16,650,000 respectively, also gives pause for thought. Aesthetically pleasing perhaps, but not indicative of refined precision. The figures should be used cautiously; not ideal, but the best we have.

The population data for individual cities was compared to the data on cities produced by the Globalization and World Cities (GAWC) Research group (GAWC, 2002). They looked at the distribution of advanced producer services across a range of cities (Taylor, 2001). They generated a data matrix of 316 cities and 100 firms in accountancy, advertising, banking/insurance, law and management consultancy. They identified firms with at least fifteen identifiable separate offices. They identified connectivity between the 316 cities. Those that had at least one-fifth of the connectivity of the most connected city, which was London, were identified as world cities. A total of 123 global cities were identified (see Taylor *et al.*, 2001 who build upon the earlier fifty-five city network used in Beaverstock *et al.*, 1999b). Again, the data are not ideal, they rarely are in the messy world of social observation and social processes, but they provide us with one of the most sophisticated sets of global city network data produced to date. The global urban network they identify is a network of and for the functioning of advanced finance capitalism. If we had chosen another set of flows, say for example, flows of migrants, remittances, drugs or weapons, a very different network would have been identified. It is the operation of advanced producer services that flow through the GAWC network. The GAWC global cities are the cities of advanced finance capitalism.

There is clearly an overlap between big cities and the GAWC global cities. The twenty largest cities cited by Brinkhoff also make the GAWC listing. Tokyo, New York, Seoul, Mexico City, Osaka and Los Angeles, to name just the largest six cities, are also GAWC world cities. Some cities are not GAWC global cities because they are too small. Advanced producer services require a significant threshold population size. Clearly then, some cities do not make the GAWC list simply because they are too small. To reduce this size effect, Table 5.1 lists only those cities that fulfilled two criteria; they had a population of more than 3 million and are not on the GAWC list. The population figure is arbitrary and different results would be produced if a different threshold were used. But, for the moment, it provides a significantly large threshold for a reasonable definition of big cities. Table 5.1 lists cities that meet the population threshold figure and are not GAWC global cities. In large measure they represent what we can term Third World urbanization: all of the cities are located in low and low- to middle-income economies, as defined by the World Bank (2001).

Table 5.1 Large non-global cities

City	Country	Population
Tehran	Iran	10,700,000
Dhaka	Bangladesh	9,950,000
Khartoum	Sudan	7,300,000
Chongqing	China	6,750,000
Kinshasa	Congo	6,150,000
Lahore	Pakistan	5,950,000
Hyderabad	India	5,850,000
St Petersburg	Russia	5,550,000
Tianjin	China	5,450,000
Nagoya	Japan	5,150,000
Baghdad	Iraq	4,950,000
Alexandria	Egypt	4,850,000
Ahmadabad	India	4,650,000
Rangoon	Myanmar	4,650,000
Wuhan	China	4,500,000
Belo Horizonte	Brazil	4,450,000
Harbin	China	4,350,000
Shenyang	China	4,350,000
Algiers	Algeria	3,950,000
Guadalajara	Mexico	3,950,000
Pusan	South Korea	3,950,000
Abidjan	Ivory Coast	3,850,000
Medellin	Colombia	3,850,000
Poona	India	3,850,000
Porto Alegre	Brazil	3,650,000
Chengdu	China	3,550,000
Monterrey	Mexico	3,550,000
Pyongyang	North Korea	3,550,000
Phoenix	USA	3,450,000
Recife	Brazil	3,450,000
Ankara	Turkey	3,400,000
Salvador	Brazil	3,200,000
Cali	Colombia	3,150,000
Chittagong	Bangladesh	3,100,000
Nanjing	China	3,050,000

A large city may not be a global city because it is sharing a national space with one or more cities that do act as a gateway for global connections. Nagoya and Alexandria are large cities without global city status that are in a national urban system that does, respectively, Tokyo and Cairo. While the connection between national and urban economies is rarely as simple as most economists would have us believe or most national statistics assume, urban global connectivity is affected by the presence of other cities in the same national space economy. In order to account for this effect, the data in Table 5.1 were filtered by removing those cities that were in national urban systems where at least one other city was identified as a GAWC global city. Table 5.2 is the result. Eleven cities were identified which met three criteria: they had a population of over 3 million, were not identified by GAWC as a global city and did not share their national territory with a global city. They ranged from Tehran with a population of 10.7 million to Chittagong with a population of 3.1 million. There are a number of reasons behind these very large cities' non-global city status. I will posit five: poverty, collapse, risk aversion, exclusion and resistance.

In large measure non-world city status reflects poverty. Some cities, despite their size, are so poor that they do not represent a market for advanced producer services. They are the black holes of advanced global capitalism, with many people but not enough affluent consumers or complex industries to support sophisticated producer services. Approximately eight of the eleven cities in Table 5.2 are located in low-income countries, and three (Tehran, Baghdad and Algiers) are in the low–medium category as defined by the World Bank. The Bank classifies countries into four categories of gross national income per capita: low, low–middle, upper middle and high. All of the cities in Table 5.2 are in the two lowest categories. The absolute values of gross national income per capita are shown in Table 5.3. These cities are in some of the poorest countries in the world. Kinshasa is the capital of a Congo whose gross national income per capita in 2000 was a pitiful $110.

Table 5.2 Black holes 1

City	Country	Population	World Bank status	Risk rating
Tehran	Iran	10,700,000	Low–medium	High
Dhaka	Bangladesh	9,950,000	Low	Significant
Khartoum	Sudan	7,300,000	Low	Very high
Kinshasa	Congo	6,500,000	Low	Extreme
Lahore	Pakistan	5,500,000	Low	High
Baghdad	Iraq	4,950,000	Low–medium	Extreme
Rangoon	Myanmar	4,650,000	Low	High
Algiers	Algeria	3,950,000	Low–medium	High
Abidjan	Ivory Coast	3,850,000	Low	Significant
Pyongyang	North Korea	3,550,000	Low	Very high
Chittagong	Bangladesh	3,100,000	Low	Significant

Table 5.3 Black holes 2

City	Gross national income per capita (US$)	Risk rating
Tehran	1,680	162
Dhaka	370	141
Khartoum	310	181
Kinshasa	110	182
Lahore	440	171
Baghdad	n/a	182
Rangoon	n/a	158
Algiers	1,580	151
Abidjan	600	141
Pyongyang	n/a	178
Chittagong	370	141

Note:
n/a = not available.

Dhaka and Chittagong are in Bangladesh, a country where the gross national income per capita was $370. In 2000 the world average was $5,240. Many of the city's population are poor, living on the margins. These cities lack a significant (consuming) middle class and an advanced urban economy. Not requiring the services of global producer service firms they are excluded from GAWC global city status. It is not legitimate to write of urbanization without globalization, since all cities partake in some form of global connections. Urbanization with only selective economic globalization is perhaps a more accurate term. The form of this globalization may be in the form of migration flows and remittances, but it is not in the form of connections between advanced producer services.

There are cases of not only endemic poverty but also cases of catastrophic decline where there has been an almost complete collapse of civil society. In recent years, Khartoum and Kinshasa, for example have witnessed the decline of the rule of law and social anarchy. War and social unrest have been the norm rather than the exception. These two cities represent cities that have internally collapsed for all intents and purposes and have been abandoned or bypassed by global capitalism. Sustained social disruption reinforces the global disconnect.

Poverty and social anarchy do not explain all the reasons why some cities are bypassed. Such global exclusion is partly a function of risk aversion by capital investors and subsequent lack of advanced producer services. There are many global metrics of risk. One private company produces a risk rating of national economies (World Markets Country Analysis, 2002). For 2002, they ranked 185 countries from insignificant risk (Luxembourg rated 1) to extreme risk (Afghanistan rated 185). The risk categories are shown in Table 5.2 and the absolute values in Table 5.3. All the cities were rated as significant, high, very or extreme risks and their numerical values were skewed towards the extreme risk end of the index. The perception of risk can be self-fulfilling. High risk deters corporations from establishing connections that in turn increases the risk rating. For our purposes the data shows one strand in a web of relationships that create the conditions for a global disconnect.

National ideologies also play a role. Tehran and Pyongyang, for example are cities where national ideologies have not encouraged global economic connections to the advanced capitalist economies. It is not accidental that the three countries for which income data was not available for 2000 were Iraq, Myanmar and North Korea. All three countries at the time were outside the global grid of international data collation exercises just as they were outside many other global discourses. Fundamentalist beliefs, of a religious and political nature, have sought to resist the encroachment of a global capitalist, because of the fear of secular beliefs in the case of Iran and of capitalist hegemony in the case of North Korea. However, cities that have severed themselves from economic globalization often find it difficult to resist the pervasive influence of cultural globalization, especially amongst the young. In the summer of 2001 it was possible to see graffiti on the walls of public buildings in Tehran, in English, lauding Madonna.

Theorization of the black holes is only at a primitive stage. Table 5.4 lists four ideal types of large, non-global cities: the poor city, the collapsed city, the excluded city and the resisting city. The table represents Baghdad before the invasion in 2003. The table lists some exemplars. There are clearly connections between these four types and in reality most of the cities listed in Table 5.4 have elements of all four in differing proportions. Social collapse tends to occur more easily in very poor cities, collapse can often induce anti-capitalist ideologies that not so much inhibit global connections as justify the lack of them. The interconnections between the four types are indicated in the table by the repetition of certain cities. Thus Kinshasa and Khartoum appears as both a poor and collapsed city, while Pyongyang is listed as both an excluded and a resisting city. Most of the cities exhibit characteristics of each of the four ideal types.

LOOSE CONNECTIONS

Cities are connected in varying degrees to the rest of the global urban network. We can use the GAWC connectivity data along with population

Table 5.4 A typology of non-global large cities

Description	Examples
Poor city	Dhaka, Kinshasa, Khartoum
Collapsed city	Kinshasa, Khartoum
Excluded city	Baghdad, Pyongyang
Resisting city	Tehran, Pyongyang

data to measure a city's degree of connectivity. Taylor *et al.* (2001) calculated a connectivity value for each of their 123 world cities based on their producer services data. The values ranged from 0.196 to 1. The most connected city, London had a value of 1, the next was New York with a value of 0.976 all the way to Lagos with a connectivity value of 0.196. Figure 5.1 plots the GAWC connectivity value of each of the 123 global cities against their population. The population figures were derived from Brinkhoff (2001) who gave the figures for all cities with more than 1 million. For cities with less than 1 million, and this included sixteen cities (Abu Dhabi, Bratislava, Calgary, Dubai, Geneva, Hamilton, Luxembourg, Manama, Nassau, Nicosia, Port Louis, Oslo, Quito, Wellington, Zagreb, Zurich), the latest population figures were taken from the respective most recent national census. Connectivity was conceptualized as a function of population size, a regression line was fitted through the data and the resulting linear regression is presented in Figure 5.1.

More sophisticated analyses could use a variety of non-linear relationships to tease out the more subtle relationships between connectivity and population caused by such general trends as the return to scale and negative externalities of large cities (see Krugman, 1991) as well as the particular feature of the GAWC data that has a small number of very highly connected cities. However, the linear regression model is a simple tool that allows us to identify, relatively easily, the position of individual cities compared to the aggregate pattern. The value of individual cities varies from this equation; the difference is referred to as a residual. Those with a connectivity greater than that predicted by the equation have positive residuals, and conversely,

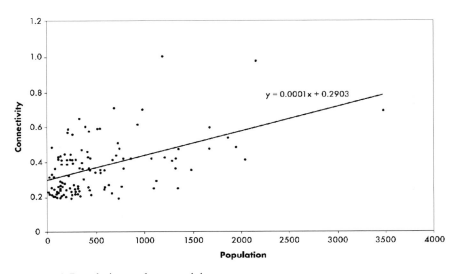

Figure 5.1 Population and connectivity.

those with a connectivity less than that predicted by the equation have negative residuals. The residuals are revealing since they indicate connectivity greater or lesser than that predicted by population size alone. The residuals are plotted in Figure 5.2 and the cities with the very largest residuals are noted. Tables 5.5 and 5.6 list the ten cities with the largest negative and positive residuals respectively.

The cities listed in Table 5.5 are GAWC global cities that have a degree of connectivity less than that predicted by their population. This is a crude measure, but nevertheless it provides a provisional look at loose connections. The least connected city relative to its population is Calcutta, followed by

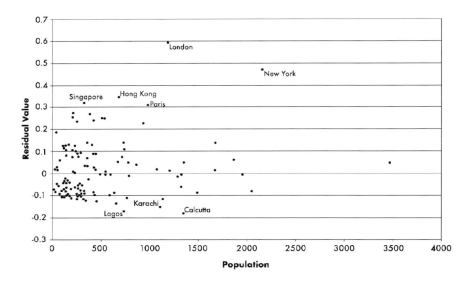

Figure 5.2 Plotting residuals.

Table 5.5 Loose connections

City	Residual
Calcutta	−0.175
Lagos	−0.165
Karachi	−0.147
Chennai	−0.131
Guangzhou	−0.121
Kiev	−0.115
Rio de Janeiro	−0.113
Pittsburgh	−0.113
Casablanca	−0.110
Lima	−0.108

Table 5.6 The well-connected cities

City	Residual
London	0.591
New York	0.470
Hong Kong	0.348
Singapore	0.322
Amsterdam	0.278
Milan	0.275
Frankfurt	0.257
Madrid	0.253
Toronto	0.251

Lagos and Karachi. From the previous discussion we would expect this pattern. The less connected cities tend to be very large cities in poor countries. Their weight of population is not matched by the corresponding amount of producer service activity found in the rest of the world. Further down the table are Kiev and Pittsburgh. The economy of Ukraine has collapsed in recent years while Pittsburgh has witnessed severe deindustrialization whilst many of the producer service functions can easily be handled by large cities close by such as New York and Philadelphia. The more connected cities, shown in Table 5.6, include London, New York, Hong Kong and Singapore and a strong representation of European cities as well as Toronto. London and New York, despite their huge populations, are more connected than any other cities, a function of their centrality in this global urban network. The remaining cities are much smaller but have significant connectivity. The results are more suggestive than definitive but they suggest cities differentially connected to a global urban hierarchy.

It is important to note in passing that the largest positive residuals are substantially larger than the largest negative residuals, which suggests that the network is dominated by a small number of exceedingly well-connected cities. The residual value for London is almost six times the size of the value for Pittsburgh for example. The results show a global urban network dominated by a few key cities. London, New York, Hong Kong and Singapore provide the major hubs of this global urban network.

The results give us empirical hints of the linkages that both embody and reflect patterns of development and underdevelopment. There is a national bias to the development literature. Simple classifications of national income and development status need to be replaced with more place specific flow data. Cities rather than national economies are the linkages in the global economy. Global urban flows are one of the most important structuration mechanisms for the development of underdevelopment.

Using similar but more sophisticated global urban network flow data may also provide us with an avenue to transcend the current demarcation in urban studies between First World and Third World perspectives. There is a geography to urban theory with First World cities more noted for connectivity and centrality while Third World cities are more cited for isolation and marginality. Producer services and finance centers are studied in First World city research while squatter settlements and the informal economy are the dominant focus of Third World city studies. Each school has a vested interest in maintaining this academic division of labor. Using global connectivity data allows us to link development issues with urbanization trends and to connect First World and Third World cities in a more global urban paradigm.

CONCLUSIONS

This chapter has been a rudimentary exploratory empirical analysis that focused attention on the missing gaps rather than the connected nodes of one particular global urban network. The data sets were not ideal and the analysis was relatively crude. The analysis was pitched at an aggregate scale and did not discuss the issue of scale. For example, global cities have marginalized populations while black holes also have transnational elites. There are a number of problematic issues: only one particular global urban network was analyzed, only connectivity and population were used to identify black holes and loose connections, and national data was combined with urban data in Tables 5.2 and 5.3 to infer causal connections. Clearly, further work is desirable. However, the chapter has provided the empirical beginnings for a debate on the uneven nature of globalization. The most startling point to emerge is to what extent this simple analysis managed to pick up some obvious examples of global disconnect such as Kinshasa, Rangoon and Pyongyang. Poverty, economic and social collapse, exclusion and resistance were posited as reasons for the existence of very large cities with few global economic connections. Four ideal types were suggested: poor city, collapsed city, excluded city and resisting city. These cities were either ignored, abandoned or excluded by global capital or were sites of resistance against capitalist incorporation.

The picture of loose connections presents a similar pattern of large poor cities in Asia and Africa much less connected than London, New York, Hong Kong and Singapore. The absolute values suggest a network dominated by a few very well-connected cities.

The work presented here has been more suggestive than definitive with the balance weighed towards more questions than answers. However, it suggests a more inclusive study of globalization and cities. Theorized case studies of less connected cities may present an intriguing opportunity to understand and explore the underside of globalization. Silences and voids are also part of globalization. It is just as important to identify the black holes and loose connections as well as the important nodes of global urban networks.

6 Tensions in the global city

Tensions between different groups in the city are neither new nor particularly restricted to processes of globalization. The city is an arena in which different groups are always seeking to advance their interests and protect their position. Sometimes there are tensions within groups. For example residents who are both users of public services and taxpayers have conflicting needs. As users of public services they want a high level of accessible and reliable public services, but as taxpayers who foot the cost of public services they would like to see limits on the amount of taxes that they pay. Similarly, within the business grouping: the concerns of big multinational corporations may not chime with the needs of local family businesses, while footloose companies may have very different place concerns to companies embedded in particular locations. The city is replete with competing interests and both evolving and disintegrating alliances. The city is riven with a series of tensions.

THE DECLINE OF THE KEYNESIAN CITY

Urban economies are always in a state of flux. Ripples in the space economy caused by new technologies, overseas competition, the rise and fall of specific industries in particular locations and changing capital–labor relations make urban economies always precariously poised between stability and change. Long periods of relative stability may be replaced by periods of rapid fluctuation. Presently, we are in a period of rapid flux. In recent years urban economies around the globe have been fundamentally restructured by a more pronounced and deepening economic globalization.

The single biggest tension in globalizing cities is caused by economic globalization and the consequent restructuring of the urban economy and local labor market. There has been a major restructuring of the global economy that in summary can be expressed as deindustrialization in the first world, increasing market penetration of the second and Third World, and industrialization in the Third World. And in all worlds there has been a deepening and widening of international capital flows and a loosening of trade barriers. The neo-liberal agenda of free trade in goods and services, no restrictions on capital flows and limited national economic sovereignty are more of a goal than a concrete reality but it is the direction that the global economy has been moving towards in the last twenty-five years.

The global shift in manufacturing employment has involved the large-scale loss of manufacturing jobs in the cities of the developing world. There has been a globalization of manufacturing production that, given the importance of variations in the costs of labor, has meant the large-scale loss of manufacturing jobs in the western economies. The result: many industrial cities in the developed world have seen the virtual destruction of their traditional economic base. The net effect in first world cities is a shift from manufacturing to service employment. This has not just been a shift in the allocation of economic activity. The manufacturing industries were the mainstay of organized male labor, the economic arm of social democratic movements and the most important element in the capital–labor agreement that constituted what I will term as the Keynesian City. This city is named after the English economist John Maynard Keynes (1883–1946) who stated that the normal operation of the market did not necessarily lead to the best outcomes. Writing in the Great Depression, Keynes argued that government had a major role to play in stimulating aggregate demand in the economy. Without the government to spend money, the market could function sub-optimally with high levels of unemployment. While Keynes was mapping out the theory of an activist government in capitalist societies, Roosevelt's New Deal was putting it into practice. Faced with massive unemployment Roosevelt's Administration, beginning in 1933, began using government spending to get the economy working at higher capacity, soak up unemployment and thus secure social stability.

The period from 1933 to the 1980s marks the high point of the Keynesian City when there was a consensus between capital and labor on the role of government. At the aggregate level, government spending was to be used to stimulate demand so that unemployment would be limited and controlled. Social harmony in capitalist society was underpinned by government commitment to softening the social consequences of downturning business cycles and government spending to fund social welfare programs that ensured that the majority of the population had access to relatively affordable health, housing, education and social welfare. There were major national differences. In the USA business interests held a stronger hand in comparison to North-West Europe where organized labor was relatively stronger and social welfare programs were not so curtailed by greater resistance to taxes and to the role of government in general. Military spending used to sustain the USA's global reach also militated against social welfare spending. But on both sides of the North Atlantic, the deal was predicated upon the economic muscle of organized labor to force concessions out of business and government. The Keynesian City was marked by the presence of organized labor not only in the population and as a dominant actor in the workforce but as a significant element in the structuring of the city. Taxes were raised to minimize unemployment and the vagaries of the market on ordinary people. There were effects on the urban structure. In Britain for example, urban housing markets had a significant amount of public housing that constituted almost 30 percent of the housing stock across the country and sometimes as much as 60 percent in specific cities. For many of its ordinary citizens, life in the Keynesian City took away the rough edges of a capitalist economy.

From the 1980s the Keynesian City began to disappear. There were many factors at work that have been widely documented, including: the persistence of stagflation that seemed to disrupt the balancing act of government spending that could minimize unemployment while avoiding inflation; growing resistance to government taxation as programs were funded by deepening and widening the income tax and local property tax base. However, what also underlay the seismic shift was the declining power of organized labor. The loss of manufacturing jobs and the consequent decline in the size and import-ance of organized labor meant that the hand of capital was strengthened. Beginning in the 1980s a new narrative takes over; limiting government spending, especially on welfare programs, reducing social subsidies, freeing up markets, globalizing economies. In both the USA and UK, for example, the Reagan administrations and the Thatcher governments embodied the political shift that took as one of its goals the further erosion of the power of organized labor. Since then there have been self-imposed limits on tax increases, a redirection of government spending and a reorientation in the nature of the city. The mantra of "small government" is just one of the slogans used to roll back government social spending.

The Keynesian City has been replaced by the Entrepreneurial City. As the central government reduces social spending and the tax base declines in many formerly buoyant industrial cities, city authorities have been more concerned with generating money than spending. And even the spending has become more orientated to creating a fertile business climate. This shift involves specific strategies such as looking at the development opportunities of plots of publicly owned land and creating public–private partnerships in which business and local governments pursue joint developments. The political discourse of the city has becomes less about the social welfare of citizens and more about the competitiveness of local business. The city is seen less as a place of residence and more a site of business. The urban debates become dominated by improving the competitiveness of local businesses and attracting footloose capital. To a certain extent the debate is pitched at the level of general welfare, a variation of the trickle-down theory, in which large developments are justified in terms of job creation and positive tax benefits. The arguments are reinforced by what we may call the TINA factor, a phrase oft used by Margaret Thatcher: there is no alternative. Coming at the end of a long period of Keynesian dominance, it is as if the center-left has run of out ideas. The moving edge has been won by the center-right who seem to have a more coherent policy and a more muscular commitment to informing and shifting public opinion.

The shift from Keynesian to Entrepreneurial City varies by country and city. Older industrial cities in the USA, have moved furthest across the spectrum, cities with more buoyant economies in countries with a stronger commitment to social welfare programs moved least. But everywhere the shift is in the one direction.

The shift from Keynesian to Entrepreneurial City also involves a shift in political culture. People as residents and citizens are reconstituted into people as workers and consumers. It is a less caring city. Being poor becomes less a condition and more of a moral failing. To be economically marginal in the entrepreneurial city is to be a social threat.

The furthest move is towards what Neil Smith (1996) describes as the revanchist city. The term comes from a political movement in France in the late nineteenth century that sought to discipline the French people. Smith sees a similarly "reactionary campaign" in US cities that "portends a vicious reaction against minorities, the working class, homeless people, the un-employed, women, gays and lesbians, immigrants." The list is so long it is difficult to see who is left out. Despite the broad-brush rhetoric, Smith has identified an important strand of a reactionary urban discourse. A more nuanced analysis would identify groups that are particularly vulnerable; for example, low-income immigrants (not all immigrants) single-parent, low-income women (not all women). Don Mitchell's (2003) work on the public space and the homeless in US cities reveals the extent of this reaction on one of the most vulnerable groups and its consequent restructuring and redefinition of urban public space.

The reactionary trend has been reinforced by the privatization of urban public space and urban public life, particularly prevalent in large US cities. The decline in communal public spaces and the rise of commodified, semi-public, semi private spaces, such as the mall, distances those less able to consume and buy. General urban public spaces became more segmented into income groups. As we move from a Keynesian City to a more privatized city, the marginal disappear as citizens and reappear as a threat. And the more people hide behind gated communities, live in segregated suburbs and patronize socially segmented sites, the more urban public space becomes less of a site of regular interaction and more of a scary encounter. The higher the walls go up, the less the feeling of safety and security. As middle and upper income groups retreat from urban public space the more it becomes a place of threat and danger to be disciplined, policed, controlled and avoided.

The decline of organized labor and the consequent decline of the Keynesian City was caused in part by the global shift in manufacturing employment from the cities of the first world to the cities of the developing world. However, while a significant feature of economic globalization in the 1970s and 1980s was the global shift in manufacturing, in more recent years there has been a significant increase in the global shift in service employment. A new round of economic globalization is sending a range of service employment including engineering design software development and even some financial services from the developed to the developing world. Back offices in Bangalore, India process home loans for US mortgage companies while many insurance claims made in the USA are routinely processed in offices situated in New Delhi. The economics are ruthlessly simple. It costs Boeing roughly $6,000 a month for aircraft engineers. Off-shore workers, even those with a master's degree, cost only $650. Software designers in the USA cost $7,000 a month while experienced designers in China and India cost only $1,000 a month. A report in *Business Week* on February 3, 2003 estimated that almost 3.3 million high-level service jobs would move off-shore from the USA. Global companies now routinely outsource work previously only done in core locations to cheaper locations. General Electric now employs 6,000 scientists in ten countries. In the 1970s and 1980s engineers would come to the USA and Europe, now the jobs come to them. In cities across the world such as Shanghai, Manila and San Jose service jobs previously done in first world cities are now becoming the making of a urban middle class. In Manila, for example, 8,000 foreign companies use workers in almost a dozen office complexes with fibre optic links. Peter Goodman (2003) describes the case of an architect in Shanghai drawing blueprints for a staircase for a Fortune 500 company in upstate New York. Architects in New York cost between $100 and $120 per hour; in China they cost only a fraction of that figure. More highly skilled work is now being shipped off-shore where rates are cheaper and the entire 24 hours in a day can now be utilized with a global labor force.

The consequences of this global shift in service employment are only now taking shape. We can make some tentative remarks in the form of questions for further study. Are we witnessing the beginnings of the end of the new middle class in first world cities? While the decline of manufacturing employment caused the decline of the working class, will the shift in service employment herald the beginning of the end of the middle class who up until now have benefited from the economic globalization? Will they now bear the costs as service jobs begin to disappear? Are we on the verge of a wholesale socio-spatial transformation that effectively destroys the working and middle class of the first world cities only to reconstitute them in new form in the cities of the developing world?

There are some who have done well out of economic globalization. The last twenty years have seen a significant increase in the wealth of the global capitalist class. The richest 10 percent of the world's population had incomes that were seventy-nine times higher than those of the poorest 10 percent in 1980: their incomes were 120 times higher in 2000 (Bergesen and Bata, 2002). The top 0.25 percent of the world's population owns as much wealth as the other 99.75 percent.

There is also the slightly larger global elite that Sklair (1997) suggests includes: transnational executives and their local affiliates; globalizing bureau-crats; globalizing politicians and professionals; consumerist elites (merchants and media). These groups are the main beneficiaries of the rise of global capitalism: they are the knowledge-rich (and asset-rich) individuals who obtain their power and influence by virtue of their pivotal role in managing the global economy. They constitute an international elite whose migration flows link together global and globalizing cities in a space of flows that reproduce their cosmopolitan interests and practices. They are mobile, global and cosmopolitan; they are the in the front of the large jumbo jets, the occupants of downtown luxury business hotels, the diners at expensive restaurants and the vacationeers at internationally renowned yet still discreet resorts. They both embody and represent globalization.

We can characterize the trends in a brief description of four people. First, there is the office cleaner who works for little more than $6 an hour in the headquarters of a transnational corporation in a global city like New York. They are not employed by the corporation directly but by a cleaning agency that employs mostly new or recent immigrants to the city. With no health benefits, an overheated housing market and little time or energy after the working day to gain new educational qualifications this worker maybe cleans the offices of globalization but does not benefit greatly from the experience. Perhaps the experience is better than remaining a peasant in El Salvador, but the perceived benefits are to their children and extended families, who receive remittances, rather to themselves.

Second, there is the blue-collar factory worker, born in 1950, who saw their employment opportunities shrink as manufacturing employment has shifted across the globe, and their wages reduced in real terms. Their families are now kept aloft by having more than one person in the household in work. They have seen welfare services reduced and the cost of education, health, and housing increase. They are struggling to keep afloat.

The software designer, a job description only arising in the last twenty years, has seen a rise in real income as their marketable skills have been turned into high wages in a company offering good benefits. They see themselves less as part of an organized workforce as the factory worker does, and more as a budding entrepreneur. However, the golden years of the 1990s have been tempered by the downturn of the early 2000s and looming over the horizon is the possibility that the work may be eventually outsourced to a skilled yet cheaper designer in Shanghai or Bangalore.

Then there is the senior executive of a transnational corporation whose income along with stock offerings has exploded since 1980. Their salary is now 500 times the average salary of factory workers in the company they manage; they have tremendous benefits and a cosmopolitan lifestyle. The whole world is their place of business, the source of their consumption patterns and their playground. These four people may live in the same global city but they inhabit very different worlds.

Globalization is not the cause of social polarization. There have always been inequalities of wealth. But what globalization does is to create major cleavages within communities that live in cities and occupy national societies.

Globalization causes a more marked rupture between the space of shared economic interests and the shared space of national society and civic society.

THE RISE OF GLOBALIZING CITY REGIONS

On the other side of the global shift another story of globalization and the city needs to be told. Its basic outline involves the opening up of national markets, increased penetration of foreign investment and the concentration of economic growth in large urban regions. Scholars have identified global and globalizing city regions (GCRs), also known as Extended Metropolitan Regions (EMRs), in which most of the urban and industrial growth has been concentrated (Scott, 2001). These new globalizing city regions are the site of multinational corporations investments and new techniques of manufacturing and centers of service industries. GCRs have a number of distinctive features. The core has specialized producer services such as banking, accountancy, legal and advertising services, so in effect, they are centers of economic transactions and surveillance and hubs in the global network of flows in producer services. Also in the core are the hotels, communications and people transport that link the country to the rest of the world. In the periphery are the more intensive land uses such as manufacturing and mass housing. In rapidly expanding GCRs an outer ring of recently urbanized countryside can also be identified, as the metropolis extends its influence across the landscape.

Three GCRS have been identified in Asia Pacific: Bangkok (11 million population), Seoul (20 million) and Jakarta (20 million) which have between 35 and 25 percent of all foreign direct investment into their respective countries and constitute between 20 percent and 40 percent of respective national gross domestic product. Let us consider GCRs in two countries, Mexico and China.

In 1978 China inaugurated more reformist economic policies that opened up their economy. The next year greater autonomy was granted to two coastal provinces, Guandong and Fujian, to create export orientated industries. Five-year plans since 1981 have stressed the economic development of the coastal districts through the encouragement of export industries. In effect, export growth was spatially concentrated. With a central government eager to attract foreign investment and multinational corporations eager to utilize cheap labor sources and tap into a huge market, the scene was set for the rapid-export orientated industrialization and urbanization of the coastal districts of China. Existing cities such as Beijing, Shanghai and Hong Kong have grown into GCRs. Table 6.1 lists the importance of these three globalizing city regions to the national economy: while they constitute less than 8 percent of the national population, they attract 73 percent of the foreign investment and produce 73 percent of all exports.

Fulong Wu (2001a, 2001b) has documented some of the urban effects of this rapid globalization, with primary reference to Shanghai. In particular he

Table 6.1 Globalizing city regions in China, 1999

GCR	Population	FDI ($)	Export ($)
Beijing	25 million	4,100 million	15,700 million
Shanghai	40 million	6,400 million	32,600 million
Hong Kong	48 million	37,300 million	167,400 million
% National total	7.5	73.0	73.1

Source: Anonymous, 2003b.

points to the growing marketization of the housing and property markets and its consequences. Beginning in 1979, existing public housing was sold. And since the early 1990s there has been a real estate boom. Capital flowed into the real estate market causing a building boom and an increase in land and housing prices. The average house price soon became approximately 10–20 times the average household income in the city. In what is a metaphor for the whole process, the urban spatial structure shifted from the mixed cellular structure of Chinese traditional cities towards the pattern of market-structured, differentiated, land use patterns found in western cities. In other areas such as health provision and employment similar patterns can be found as marketization/globalization lead to a decline of equity considerations and a growing emphasis on market competitiveness. The Maoist City has been replaced by the Entrepreneurial City.

The city of Shenzhen, on the mainland section of the Hong Kong GCR, was one of the first areas of China to be opened up. In 1978 it was declared a special economic zone and tax breaks and investment incentives were implemented to encourage foreign capital. Two year later the city was nick-named, "Window to the Outside World." Growth rates were impressive and in 1990 the Shenzhen Stock Exchange opened. The city has experienced tremendous growth. The population of 300,000 in 1980 has now grown to over 4 million. Many of the industrial workers are young women; the average age of a city resident is only 29. The factories rely on young female workers. There are so many job opportunities for women the city is known as "women's paradise." Peter Hessler (2001) recounts the story of Ma Li who left her hometown in Sichuan province, after graduating from a teachers' college, to work in a factory owned by a Taiwanese businessman that made costume jewelry for the export market. As an entry worker she made 870 yuan (US$105) a month while her college classmates were earning US$40. She lived in a dormitory with six workers to a room. Ma Li remitted some of her earnings to her pay for her brother's schooling. In Hessler's telling of her story, Ma Li becomes more worldly, she is protective of her fellow workers, shielding them from the sexual harassment of the boss and moves into a three-room apartment with a male partner. Her salary rose to US$240 a month but she was becoming discontented with the long hours. When she left the job to work as a nursery school teacher with similar salaries and fewer restrictions on personal freedom her old boss told her, "You've changed. You used to be obedient." She replied, "I didn't change. I just got to know you better." Individual stories are rarely capable of bearing the full metaphorical weight of broad-scale social processes, but the concluding paragraph of Ma Li's story is worth repeating:

> I looked at Ma Li and realized that . . . since coming to Shenzhen, she had found a job, left it, and found another job. She had fallen in love and broken curfew. She had sent a death threat to a factory owner, and she had stood up to her boss. She was twenty-four years old. She was doing fine.
>
> (Hessler, 2001: 119)

A more primate urban hierarchy is found in Mexico where the capital city dominates the economy and polity of the country. Christian Parnreiter (2002) has analyzed the global integration of Mexico City in some depth. He points to the increased global economic integration of the Mexican economy since it joined the forerunner of the World Trade Organization in 1986 and the North American Free Trade Agreement in 1994. Foreign trade and foreign investment have increased dramatically: foreign trade as share of gross domestic product increased from 20 percent in 1980 to almost 60 percent in 2000. Unlike the case of China the number of manufacturing jobs in the greater Mexico City GCR actually fell in both relative and absolute terms as manufacturing employment shifted to the northern region bordering the USA. More crucial to the urban economy was service employment, especially in the finance and real estate sector. The banking sector grew on average 8 percent per annum from 1985 to 2000. However this growing sector did little to raise average wages. Average incomes fell throughout Mexico in the period since 1980, losing almost a half of their real purchasing power.

The policy shift away from import substitution to concentrating on foreign markets resulted in many negative impacts on the urban economy. The traditional center of the manufacturing sector in Mexico City, previously protected, was now opened to intense foreign competition. Wages fell and jobs were lost in the city region. Jobs moved to the border regions. The upside of global integration was that Mexico City became the preferred location of the national headquarters of large transnational companies. The city was the epicenter of foreign investment taking up 60 percent of all foreign direct investment during the 1990s. Mexico City was gradually transformed from a manufacturing city to the global gateway for a Mexico now more fully integrated into a global economy. The result was the growth of high-paying jobs in the advanced producer services, the loss of manufacturing employment and a tight squeeze placed on the lowest income groups. The global reorientation of Mexico's economy had marked effects on the capital city region: a few of the most globally connected did well and there was the slow emergence of a nascent middle class, while the majority are worse off now than they were in 1980 before the push towards global integration.

The effects of globalization rippled through the national urban hierarchy. Further north along the border region cities such as Ciudad Juarez witnessed a massive increase in population as people flocked to the new employment opportunities in the manufacturing sector. This region had been something of an economic backwater in the national space economy until the Border Industrialization Program (BIP) was established in 1966 as a response to the termination of a migrant worker program with the USA that had been in operation since 1942. The action of the US government in 1964 caused many workers to return to the northern border region. It was the BIP that created the *maquiladora* plant program along the border region to generate employment. The *maquiladora* program allowed the free import and export of raw materials and goods between the USA and Mexico for factories along a narrow border region. The factories on the Mexican side became known as *maquiladoras*. They were established along the border to take advantage of the signing of the cheaper Mexican labor rates but with easy access to US consumers, using Mexican labor to produce goods for the US market. The NAFTA in 1994 reinforced the manufacturing growth in border cities such as Ciudad Juarez, which increased its population from 122,000 in 1950 to over a million in 2000. In 1998 almost a quarter of a millon people worked in *maquiladoras* in the city. The Ford Motor Company alone, for example, had ten plants in 1998 employing over 11,000 workers. Almost one out of every two workers is employed in manufacturing plants geared for the North American and global markets. The city has 300 manufacturing plants. This very rapid industrialization has resulted in sprawling squatter settlements along the periphery of the city as population growth outpaced the city's ability to meet welfare needs (Anonymous, 2003a). With a more recent downturn in the US economy the *maquiladoras* have begun to lay off workers. Since 2000 200,000 workers have been laid off in the border region (Thompson, 2001). And even the hourly wages of between $4 and $5 a day are eclipsed by even lower waged workers in El Salvador ($1.59 a day), Dominican Republic ($1.53 a day) and China (53 cents a day). The global economy is relentless in its pursuit of cheap labor for manufacturing production, and whole complexes such as those around Ciudad Juarez may be short-lived as their comparative advantage rapidly appears, only to quickly wane.

Globalization has restructured the Mexican national space economy. One effect is the creation of manufacturing cities especially in the northern region bordering the USA. The main GCR, Mexico City has witnessed a steady deindustrialization as the city's economy has shifted from manufacturing to services. The net effects have been a growing spatial inequality between globalizing and less globalizing parts of the national space economy and growing polarization within cities between, on the one hand, those benefiting from globalization, including advanced producer service workers in Mexico

City and manufacturing workers in such places as Ciudad Juarez, and on the other hand, large sections of the industrial labor force in Mexico City.

Unfettered capitalism rewards winners and punishes losers. Globalization exacerbates this trend which is even more reinforced in the GCRs with the shift from closed to more open economies. The form of the opening varies from planned economies in the case of China and former centrally planned economies in former communist-controlled states, to import substitution economic strategies in the case of Mexico and many other developing countries toward a more direct integration into the global economy. The restructuring involves a social reorganization of the national space economy and growing inequalities between globally successful and less successful sectors and workers. Those in the upper echelons of advanced producer services, such as investment banking, now operate on a more global metric of remuneration while those in traditional sectors see the disappearance of their safety net and a decline in their job security. In the short term there is a growing tension between those profiting from globalization and those being punished by globalization. Some economic theorists would argue that while this may be true in the short to medium term, in the long term globalization will promote greater economic efficiency and ultimately greater income growth for everyone. However, we tend to live our lives in the short to medium term. In the long term, as Keynes reminded us, the one true certainty is that we are all dead.

The tensions in global and globalizing cities have been reinforced by the decline of the Keynesian City in the West, the marketization of the planned economies in the East and the reduction of the social wage, small to begin with, in the South. It is not economic globalization per se that is causing the social tensions within cities, it is the governmental responses to economic globalization. The studies that have measured social polarization and income distribution in different cities show that the range varies according to how the national society responds to market inequalities (Fainstein, 2001). Some societies defuse economic inequalities while an increasing number are reinforcing them. The city is the arena in which this tension is being played out in the starkest of forms.

7 The modalities of the global city

You arrive at an international airport, any international airport, for the first time. It is both new and strange, surroundings never before visited, and yet strangely familiar. In some sense, if you have been to one big modern international airport, you have been to them all: the similar configurations, the shared décor. The sameness is not just an accident of shared technologies or similar architects, it is a desired end. Global cities, like their airports, are all very different, yet the same. There are recurring features that exist in global cities and are the focus of desire for globalizing cities. Four of the modalities of the global city are the subject of this chapter.

GLOBAL CONNECTIONS: PORT, RAILWAY STATION AND AIRPORT

An international airport is a prerequisite for being a global city. And in this case, size does matter. A big international airport with numerous airlines and many destinations is a sure sign of global status. The international airport is the latest in a long line of important transport connections vital to global city status. The evolution of these hubs is one way of tracing the trajectory of global cities. Housing the hubs of global flows is an important cause and effect of global city status.

In the past, when most goods and people moved by sea, having good port facilities was of primary concern. While the movement of people has shifted away from sea to air transport, most goods continue to move by sea. However, the containerization of sea transportation has moved port facilities to large deep-water sites. Few city harbors have deep enough waters to cope with the large ships that carry the ubiquitous box containers of global trade. Port facilities have moved downstream to deeper-water sites. The docklands of global cities are now the site of various renewal schemes. Take the case of Baltimore in the USA and Docklands in London.

The city of Baltimore grew up around a port that was founded in 1729. The relatively small ships of the time could sail right into the inner harbor to the heart of the city. By the early twentieth century, the city was an industrial powerhouse. But after the Second World War, Baltimore began its long industrial decline as companies closed, relocated or shed thousands of jobs. The deindustrialization of the city caused major job losses. The port was losing its function as the bigger commercial ships were docking elsewhere. By the 1950s the inner harbor had become an abandoned space. The business elite responded by setting up two committees, the Committee for Downtown and the Greater Baltimore Committee, that in 1956 produced an urban renewal plan for the city. In 1964 the City Planning Council outlined a $260 million plan to redevelop the 230 acres of warehouses, sheds, docks and derelict spaces in the inner harbor. A proposed promenade along the waterfront promised public access to the shoreline. A voter-approved loan of $12 million and a federal grant of $22 million gave the fiscal jumpstart to the plan's implementation. By the late 1970s the focus of the redevelopment was on tourist events and sites. A cluster of developments built between 1977 and 1981 made the inner harbor a festival setting: the World Trade Center and Maryland Science Center were built with state and federal funding, the Convention Center funded with $35 million from state funds attracted business meetings; Harborplace, built by the Rouse Corporation, provided retail and restaurants in two large pavilions; the National Aquarium was built with $21 million from the city council; the Hyatt Hotel group, after being given a $12 million public grant, built a downtown anchor hotel.

The Baltimore plan was unusual in that it stressed pedestrian accessibility. Today it appears an obvious plan to make and implement, but at the time, in the USA, the emphasis was on providing indoor malls and catering to the car driver not the pedestrian. The Baltimore scheme was radical in that it was outdoors, downtown and was pedestrian-orientated rather than car-friendly. The plan also recycled some buildings rather than simply demolishing everything. The old Baltimore Gas and Electric Company Power Plant, built in 1916, was retained and now houses a bookstore, Hard Rock café and an ESPN outlet.

Baltimore's Inner Harbor is considered a success. The Aquarium receives over a million visitors each year, the Harborplace has proved a popular destination for visitors and citizens alike for eating out and shopping. Nothing succeeds like success. The Inner Harbor has been used as a model for other cities eager to revalue their downtown. When the city of Sydney, Australia wanted to redevelop a former dock area, Darling Harbour, close to the city center they used the Baltimore model and even hired Rouse as consultants. The result was a harborside remarkably similar to Harborplace with the festival setting of Darling Harbour almost a carbon copy of the Inner Harbor, and it has proved equally successful.

Inner harborfication is an awkward name that embodies a popular practice. It also has another meaning. While the Inner Harbor has proved a success, it has failed to ignite the overall economy of the city that has continued to lose the high-paying jobs of the industrial era. Baltimore continues to lose population and its public school system is notoriously under-funded. Behind the glitz of the Inner Harbor is a deeply troubled city.

In London the development of the Docklands followed a slightly different turn. The 16-square-mile area of Docklands was the old commercial water frontage of London. Ships sailed up the Thames to load and unload goods and people. By the late 1960s the docks were no longer to able to deal with the large container ships; the port facilities moved downstream to Tilbury. London Docklands was situated very close to the financial center known as the City. Here was vacant space beside one of the largest financial centers in the world. In 1981 a London Docklands Development Corporation (LDDC) was established to "develop" Docklands. The areas was home to working-class communities of almost 40,000. However, the LDDC effectively over-rode local opinion and laid the basis for commercial redevelopment and up-market housing. As in Baltimore, the private investment was underwritten by public investment, including the construction of a light railway system that effectively linked Docklands, a previously unconnected part of London, to the rest of the urban region and especially to the City. By 2000 almost 2,000 square meters of offices were constructed and over 20,000 houses. The area was transformed from a working-class docks to a commercial office region and an expensive residential area.

From the mid-nineteenth century railways became an important form of transport. Railway stations were important hubs of a global city. The construction of railway lines and stations also became a major form of urban disruption and restructuring, displacing thousands of people in their wake. The drive for a central location, especially in the older cities, meant that lines had to be carved through existing residential and commercial areas.

With the relative decline of the railway has come the possibility for global cities to restructure the rail lines and their stations. In Barcelona, for example, railway lines were part of the barrier that reduced public access to the shoreline. As part of the urban renewal associated with the 1992 Olympic Games, railway lines were removed and public access was improved to the shoreline and city beaches. Walking down from the Ramblas, the journey now no longer comes to a dead end for pedestrians, but continues into a dock area now turned into a festival setting that looks eerily like Inner Harbor and Harbourside.

In Paris, a railway station, the Gare d'Orsay was turned into the Musée d'Orsay. The transformation is interesting as it combines many features of global cities. The land for the railway station was ceded by the government to a railroad company so that a railway station would be built for the 1900 World's Fair. These fairs were the nineteenth-century equivalent of the Summer Olympics, large urban spectaculars that put a city and a nation on the international map. From 1900 to 1939 the station was an important link in the national railway system. However, the short length of the platforms meant that the larger trains could not use the station effectively. It then fell out of railway use and was scheduled for demolition but was saved by the postmodern turn in architecture and a revival of interest in old structures. In 1977 the French government designated it as the Musée d'Orsay and it opened to the public in 1986. As the home to many of the most famous Impressionist paintings, it has proved very popular, drawing in almost two and half million visitors every year. The former railway station has become a cathedral to modern art. Around the world, there are a growing number of examples of similar transformations.

Airports differ from both ports and railway stations in many ways but one of the most important it is their locational constraints. While global cities need close access to airports they cannot be too close. Airports take up large areas of space and generate noise and what economists term "many negative externalities." The locational trick is to maximize the positive and minimize the negative: too close and city residents are bombarded with noise and pollution; too far and travel times are increased. The siting of airports for global and globalizing cities highlights the positive externalities of global flows and the negative externalities of hosting those flows.

Let us consider the case of Washington DC. When National Airport, now Ronald Reagan Washington National Airport, was opened in 1941 it was a relatively small regional airport. The need for an international airport was noted officially as early as 1950 and in 1958 a 10,000-acre site was selected 25 miles west of the city in a relatively rural part of Virginia. The airport was the first one designed for contemporary commercial jets. Eero Saarinen, a signature architect of the time, designed the terminal building, control tower and service buildings. The terminal building won an architectural award in 1966. Initially the airport had two runways each 11,500 feet long. To link the airport to the city, 16 miles of highway were constructed to link up with interstate highways and downtown Washington. It took approximately 30 minutes to travel from the airport to the downtown area, assuming no traffic, an assumption that would become increasingly heroic as development expansion added extra traffic. The airport opened in 1962 to accommodate 6 million passengers a year. By 2000, it handled over 20 million passengers a year. Future runway increases are planned so that the airport will eventually handle 55 million passengers a year.

While the airport generates 15,000 direct jobs it also generates noise and pollution. Almost a million people are regularly affected by the noise of aircraft taking off and landing 24 hours a day. The average landing and taking-off noise for residents is 75 decibels, about the sound of a fast food blender at 4 feet. Jet fuels and deicing chemicals have leached into the local groundwater and river systems. Traffic congestion is made worse by a very poor public transport system: there is no mass transit system that directly links the metro transit system of Washington to the airport. The corridor from the city to the airport has also become a major growth node of the region; shiny office towers line the route and thousands of new homes have been built on what was only recently farmland. Noise, pollution and traffic congestion jostle with international connectivity and economic growth in the complicated mix of positive and negative externalities that make up the nature of international airports.

We can make a distinction between the early days of global city airports and the present day. In the past, airport construction was often seen as a good thing. It provided jobs and international cachet. But as the aircraft got bigger, more frequently the environmental costs became more obvious. The struggles to locate new airports highlight the social struggles.

A Marine Corp air station called El Toro was situated in the fast-growing southern region of Orange County of Southern California. The base was closed in 1998 and the area designated for commercial aviation use. The initial plan of the county, entitled *Global Gateway*, foresaw 38 million passengers a year moving through a new international airport. In 1998 a coalition of southern cities most impacted by the flight path called for an alternative plan to develop the land for commercial, residential and public space use. The city of Irvine proposed annexing the site and making a great park to rival Central Park in New York City. After intense political activity, in which county plans for the airport were proposed with downward-revised passenger numbers, county supervisors supported Irvine's annexation and in 2003 the proposed airport plan was abandoned. A well-organized affluent community had successfully resisted a new airport by arguing against its construction because of its deleterious environmental effects. The noise pollution and possible health and hearing impairment were all cited as reasons. And a powerful self-interest argument was that noise, pollution and traffic congestion would drive down house prices. In contrast, turning the 4,500-acre site into a park and nature preserve would only increase property values. To some extent it is an example of NIMBY (not in my backyard); there are international airports in the greater metropolitan region, and residents were not saying no to international airports, just no to having them in their backyard. However, the point is that forty years ago such debates would not have gained such political traction, as global airports were then a much easier sell. The El Toro saga shows how much the negative environmental externalities of airports have become an important part of the debate on international airports.

Even with existing airports community resistance has stiffened. In San Francisco, for example, organized groups have challenged plans for the extension of runways into the bay by pointing to the fiscal costs of over $3 billion and the deleterious environmental costs. Around the world new runways and extensions to existing runways generate costs and resistance. Airports are not just nodes in the global network of flows; they are sites of major environmental impact that highlight the tension between international connectivity and local livability.

GLOBAL SPECTACLES: SIGNATURE ARCHITECTS AND COSMOPOLITAN URBAN SEMIOTICS

Globalization involves the creation of a global system of signs and meanings. One of the most recognizable global languages is architecture, a commercial art form that turns visions into concrete realities and solidifies messages of power and prestige. Global and globalizing cities need and want the signature buildings of famous architects as they give a sense of cultural seriousness, a feeling of competing in the global arena at the highest level.

The term signature architect has been used to refer to a small group of well-known architects whose very name has an aura of architectural prestige. They are hired to design the set pieces of architectural spectacle, the prestigious corporate offices, the high-profile cultural ensembles of art galleries and opera houses.

The winners of the Pritzker architectural prize provide a roll call of some of the most well-known signature architects. The prize was established in 1979 by the Hyatt Foundation to honor a living architect. The winner receives a $100,000 cash award as well as a bronze medallion and international recognition. The prize is open to living architects and, so far, there have been 500 nominations from forty-seven different countries. Like all prestigious prizes, such as the Nobel and the Macarthur, politics plays a more significant role than simple quality. Who gets the prize is as much a function of connections as architectural worth. Many good architects have not received the prize, while the work of some who *have* may not stand up to sustained critical scrutiny. A significant absence is that of Richard Rogers whose buildings include the Pompidou Center (with Renzo Piano), the Lloyd's Building and the Millennium Dome in London. However, the prize-winners provide a useful listing of signature architects (see Table 7.1). It is an international group, albeit the sort of restricted internationalism that favors the rich countries of the world. Seven of the winners are from the USA, twelve from Western Europe, three from Japan and one from Australia. By 2003 there were only two from non-rich western countries, Brazil and Mexico: international perhaps, but not truly global.

The work of these signature architects falls into a number of distinct categories. First there are the corporate headquarters. The bigger the company the bigger the name needed to convey economic heft and cultural seriousness. Philip Johnson's AT&T building in New York City not only housed the headquarters of the economic giant it also was one of the first major postmodern buildings. Bank headquarters have been a good source of work for big-name architects. Their signature on a bank building gives the right amount of class and dignity to an organization taking entrusted with the care of other people's money. I. M. Pei designed the Bank of China building in Hong Kong and Norman Foster did the Commerzbank headquarters in Frankfurt as well as the Hong Kong and Shanghai Bank in Hong Kong. Second, there are the cultural ensembles of art galleries, symphony halls and opera houses that tell of a city's cultural importance. Examples include I. M. Pei's Meyerson Symphony Center in Dallas and his East Building of the National Gallery of Art in Washington, James Stirling's Art Gallery in Stuttgart, Frank Gehry's Guggenheim Museum in Bilbao, and Rogers and Piano's Pompidou Center in Paris. Municipalities will pay

Table 7.1 The Pritzker Prize winners

1979	Philip Johnson (USA)
1980	Luis Barragan (Mexico)
1981	James Stirling (UK)
1982	Kevin Roche (USA)
1983	I. M. Pei (USA)
1984	Richard Meier (USA)
1985	Hans Hollein (Austria)
1986	Gottfried Boehm (Germany)
1987	Kenzo Tange (Japan)
1988	Gordon Bunshaft (USA) and Oscar Niemeyer (Brazil)
1989	Frank Gehry (USA)
1990	Aldo Rossi (Italy)
1991	Robert Venturi (USA)
1992	Alvaro Siza (Portugal)
1993	Fumihiko Maki (Japan)
1994	Christian de Portzamparc (France)
1995	Tadao Ando (Japan)
1996	Rafael Moneo (Spain)
1997	Sverre Fehn (Norway)
1998	Renzo Piano (Italy)
1999	Norman Foster (UK)
2000	Rem Koolhaas (The Netherlands)
2001	Jacques Herzog and Pierre de Meuron (Switzerland)
2002	Glenn Murcutt (Australia)
2003	Joern Utzon (Denmark)

the premium required for signature architects in order to give prestige to their projects, architectural respectability to their buildings and global recognition to their cities. The new German parliament building, the Reichstag, in Berlin was designed by Norman Foster, an air terminal in Osaka was designed by Renzo Piano and the Public Library in Seattle was designed by Rem Koolhaas. Signature architects are part of the globalizing projects of cities.

While the winners were from a narrow range of countries, their work is international. If we take just one of the signature architects we can see the range of their work. Norman Foster, who won the prize in 1999, has designed a range of buildings (see Table 7.2). Foster's work, which includes corporate headquarters, cultural ensembles and airports in cities around the world, constitutes the experience of most signature architects. Beginning in places like Ipswich and Norwich in the 1970s his work is now found around the world in Berlin, Frankfurt, Hong Kong, London, Nîmes and Palo Alto. Even the offices of signature architects are international. Foster's practice employs 500 people worldwide in three main offices in London, Singapore and Berlin. The work of such signature architects is designed for global and globalizing cities, ambitious foundations and rich patrons. Their work both reflects and signifies global status and globalizing tendencies on the part of corporations, foundations and cities.

Table 7.2 Selected buildings of Norman Foster

British Museum Great Court, London, England, 2001
Carré d'Art, Nîmes, France, 1984 to 1993
Center for Clinical Sciences Research, Palo Alto, California, 1995 to 2000
Chep Lap Kok Airport, Hong Kong, 1992 to 1998.
Commerzbank Headquarters, Frankfurt, Germany, 1991 to 1997
Hong Kong and Shanghai Bank, Hong Kong, 1979 to 1986
IBM Pilot Head Office, Cosham, England, 1970 to 1971
Joslyn Art Museum Addition, Omaha, Nebraska, 1992 to 1994
London City Hall, London, England, 2003
London Millennium Bridge, London, England, 1996 to 2000
New German Parliament, Berlin, Germany, 1992 to 1999
Renault Distribution Centre, Swindon, England, 1980 to 1982
Sainsbury Centre, Norwich, England, 1977
Stansted Airport, London, England, 1981 to 1991
Willis Faber and Dumas Headquarters, Ipswich, England, 1970 to 1974

For the global and globalizing cities cultural ensembles play a significant role. Art galleries and symphony halls are important parts of the cultural economy of a city and of the global semiotics of cities: they tell of a city that has moved beyond mere money-making. Any city with truly global pretensions needs art and culture. It is part of the atmosphere of cosmopolitanism that both signifies and creates global cities. We can look at three examples that involve some of the cited signature architects; the Sydney Opera House, the Pompidou Center, and the Guggenheim Museum in Bilbao.

The Sydney Opera House sits on a site in the middle of a great harbor. It was one of the first sites of European settlement in Australia. By the early 1950s, the wonderful setting had the prosaic job of housing a tram shed. The New South Wales state announced a design competition in 1955. A little-known Danish architect, Joern Utzon, submitted the wining design. It pictured a wonderful complex of white roofs that sat like billowing sails in the middle of the harbor. Construction began in 1959 and after the usual delays and cost overruns the building was completed in 1967. It was and remains one of the great buildings of the world. I have seen it on many different occasions and it never fails to evoke wonder and delight. It is one of the most successful cultural ensembles of any of the globalizing cities. It gave Sydney an immediate icon, widely recognized around the world; its outline was used as part of the logo of the 2000 Summer Olympics. It has become one of the instantly recognizable features that globally signify the city. The other major city in Australia, Melbourne, has been less successful despite many efforts to replicate Sydney's success. There are many reasons behind the shift from Melbourne to Sydney in terms of major global gateway city, but the role of the Sydney Opera House should not be underestimated. Great buildings may not only signify globalization, but they play an important part in the process.

The Pompidou Center emerged from an international competition held by the French government to build a temple devoted to art. There was the perennial French need to play the role of cultural heavyweight. Lacking an empire and losing economic dominance, French political discourse over the past fifty years has tried to map out a distinctly French global position in areas like culture where they feel they have global edge. The competition was announced in 1970. The winning design by Renzo Piano and Richard Rogers was a bold innovative design that did not have the elegant façades that was standard for most public art buildings. The design exposed the working parts of the building such as ducts and pipes across the surface of the building. Construction began in 1972. Initially called the Centre Beaubourg, a name most Parisians still use, the building was later named after the initiating President of France, George Pompidou. The center was also an act of urban renewal built in one of the city's oldest districts, the 4th arrondissement, for some, a shabby part of town, for others a place of character where eighteenth-century buildings were sacrificed to gentrifying trends. The 50,000-square-foot building opened in 1977 and has become an important center attracting millions of visitors each year. The initial building cost around US$200 million and a major renovation in 1996–2000 added another US$80 million to the final cost. The Pompidou Center was just the beginning of a major set of prestige buildings in the French capital. The Opéra Bastille, the National Library, La Grande Arche as well as I. M. Pei's glass pyramid in the Louvre were major elements of the *grandes ensembles* vision of the two-term president François Mitterrand who served from 1981 to 1996. His architectural legacy was designed to reflect Paris's dual role as the capital of France and its position as France's global city.

The Guggenheim Museum in Bilbao is now one of the most-cited examples of contemporary architecture. Frank Gehrys' voluptuous titanium walls joined the roll call of truly important buildings. Two forces were at work in producing this icon. On the one hand there was the Basque government that wanted to create an economic alternative to the city's traditional emphasis on shipbuilding and manufacturing. Global shift had undercut the city's economic base and in the 1980s a new plan was introduced for the city that stressed tourism and cultural economics. A new museum was seen as an important element in the creation of a new economic growth machine for the city. On the other hand, there was the Guggenheim Foundation, which under the directorship of Thomas Krens wanted to embark on an ambitious international branding of the Foundation. Eventually Guggenheim Museums were built in Las Vegas, Venice and Berlin and more were planned in Russia and South America. The Guggenheim name was an important commodity that could be parlayed into new museums in negotiations with municipal and state governments. In 1991 Basque government officials approached the Guggenheim Foundation and the next year a deal was struck for a new museum on an industrial site in a former warehouse district of the city. It was an opportunity to put Bilbao on the map, attract tourists and gain global recognition. The museum was part of an ambitious rebuilding of the city including a subway system designed by Norman Foster and waterfront developments by Cesar Pelli. The two clients, the Basque government and the Guggenheim Foundation got lucky. The architect Frank Gehry produced an architectural marvel. The building was completed in 1997. Thin titanium was wrapped around steel frames to produce a sensual set of curves that covered a column-free space 1,340 meters long and 30 meters wide. The building was a great success with both the architectural establishment and the general public. Between 1.5 and 2 million people visited the museum in the first year and it has continued to attract tourists. In an assessment of the museum Beatriz Plaza (2000) presents a convincing case that it increased tourism to the city by over 50 percent. It was a winning combination of architectural innovation and urban renewal that helped to generate tourist money and global recognition. Nothing succeeds like success and other cities began copying the template of big-name architects and cultural ensembles with a renewed vigor.

Large-scale cultural artifacts are not a new phenomenon; the Eiffel Tower is a nineteenth-century example, but given the global stakes, the competition has increased in the use of unique cultural artifacts as both urban economic multipliers and cultural markers of distinction. We can now gauge the global pretensions of a city by the scale and number of signature architects' designs for cultural artifacts and ensembles. Los Angeles has seen two major developments in recent years, the new Getty Center designed by Richard Meir that opened in 1998 and the Walt Disney Concert Hall, designed by Frank Gehry that opened in late 2003. Both meet specific cultural needs for art and music but also fill a perceived cultural void for a wannabe global city. The hunger for respect is evident in the remarks of the Director of the Getty who said that, "It will make it easier for serious people to persuade themselves they might come and live in Los Angeles." A press release for the Disney Concert Hall noted that it was a "work of art that will enhance the city's cultural-architectural landscape."

Architecture is a form of fashion. Styles can soon age and what was considered cool may soon turn into boring. Cities are always on the look-out for the next talent, the next innovation that will give the edge of the present turning into the future, rather than a replication of the past. Hot new architects, possible Pritzker winners of the future, include Zaha Hadid who designed the Rosenthal Center for Contemporary Art in Cincinnati. The other finalists for the commission were Bernard Tschumi and Daniel Libeskind who has moved from relative obscurity to recognition, at least in the eyes of the general public, because of his winning design for the new World Trade Center in New York City.

High-profile building projects spectacularize the urban and raise the value of the city in the commodified images of place that flood the world. The spectacular building by the signature architect is all part of the stage-setting for an urban globality. Global cities are not just economic command centers or nodes in a network of global flows; they are sites and setting for the global spectaculars. Urban globalism needs to be performed, enacted and witnessed.

There is also an urban semiotics of cosmopolitanism embodied in the rash of residential developments occurring in globalizing cities in developing countries. Take the case of Beijing where an American neo-traditional-type residential area, named Orange County, just to drive home the message, is the latest in luxury developments aimed at the newly wealthy's taste for US and European style of residential accommodation and suburban developments (Anonymous, 2003c). "International" has become the desired style of many residential complexes sprouting around the globalizing cities in the developing world. Developers are quite literally building cityscapes that concretize global influences. The new "international style" of residential embodies the active appropriation of global forces to local housing markets. These luxury complexes are the site of a complex set of processes including the commodification of the housing market, the emergence of an affluent class with more global tastes, the transplanting of cityscape designs and the local use of international styles to make the local more connected to the images and symbols of the global rich. A global culture is practiced in these architectural forms with the goal to "create an equivalence between the residential developments and the wealthiest locations in the world" (King, 2002: 86).

GLOBAL CULTURES AND THE COSMOPOLITANISM OF THE GLOBAL CITY

Globalism is assigned to cities not just on the basis of their economic transactions or their transport connections; culture, in a variety of guises, plays a role. Raymond Williams warned that culture is one of the two or three most complicated words in the English language. He then went on to unpack some of the rich variety of meanings embodied in the term, including the notion of tending to something as well the idea of intellectual and artistic activity (Williams, 1976: 76–82). This double definition is useful here because it contains the notion that culture is cultivated and managed as well as signifying the practice of the arts and the production of aesthetics.

Culture has been managed in globalizing cities. In the late 1980s and 1990s the government of Singapore began an aggressive campaign to make the city-state a "global city of the arts." A cabinet minister was made responsible for the arts, a national gallery was renovated and $500 million was made available to build a world-class arts center. Economic forces prompted the attention to the arts. A government report argued that cultural and entertainment industries were economic activities in their own right, that they enhanced Singapore as a tourist destination, improved the quality of life and helped people to be more productive, and contributed to a vibrant cultural and entertainment scene that would make Singapore more interesting for foreign professionals and skilled workers and could help attract them to work and develop their careers in Singapore.

(Kong and Yeoh, 2003: 176)

Culture, in other words, was an economic sector to be encouraged as a form of economic diversification, a vital requirement to attracting the global elite and an important element in the construction of the global city. In 1991 the government of Singapore formally sought to develop a "world class city." The ambitious plan called for the usual elements of a city attuned to business, but it recognized the need for greater opportunities for leisure and cultural pursuits in order to propel the city to the next level of the global urban hierarchy. A government minister noted that "we need a strong development of the arts to help make Singapore one of the major hub cities of the world" (quoted in Kong and Yeoh, 2003: 179). The government provided the necessary hardware in the form of improving and building cultural facilities including the National Theater and the Esplanade, a custom-designed building that cost S$667 million and included a concert hall and theater, specifically built to host "world class" acts.

Two important themes in globalizing cities are the capitalization of culture and the globalization of culture. Culture has been capitalized. A variety of cultural industries can be identified, including performing arts, film, music, sports, museums and galleries. In the international division of labor there are some cities, such as Los Angeles and Bombay, that specialize in specific cultural industries. Hollywood and Bollywood are centers of important film industries. Globalizing cities actively encourage cultural industries as an important economic sector in their own right and as a way to cosmopolitanize the city.

This is not a new phenomenon. A character in Sinclair Lewis's 1922 novel *Babbitt* addressed a meeting of the city's booster club with the words:

Culture has become a necessary adornment and advertisement for a city today as pavements or bank-clearances. It's Culture, in theaters and art-galleries and so on, that bring thousands of visitors to New York every year and, to be frank, for all our splendid attainments we haven't got the Culture of a New York or Chicago or Boston. The thing to do then, as a live bunch of go-getters, is to capitalize Culture; to go right out and grab it.

Go-getters in cities around the world are heeding Babbitt's message. The relatively more fluid nature of the global urban hierarchy is providing enough space as well as incentive to make it worth the while of cities to seek to capitalize culture. And once the global race begins, the stakes are raised as successive rounds of cultural investment produce subsequently higher rounds as the competition increases.

The crude distinction between culture and economics no longer holds. The idea that culture sits airily above the mundane economy is no longer tenable. Cultural industries such as film, television, sports and art exhibitions provide jobs as well as generate income. Cultural industries are an important part of economic diversification necessary to ensure the long-term economic success of a city in the global economy. The political economy of culture also operates in more subtle ways. A certain cultural richness is assumed to be the mark of a global city. There is an assumed level of cultural capital necessary to maintain a city's competitive position in the attraction of capital, tourism and skilled labor. The distinction of high culture of opera, visual arts and symphonic music considered vital in the mid-twentieth century has been leavened with the more popular cultural attractions such as film, sports, popular music and ethnic cuisines. This range of cultural capital feeds directly into the creation of a cosmopolitan sensitivity vital to maintaining a global competitiveness. As the Minister for Information of Singapore noted in 1991:

> we also need the arts to help us produce goods and services which are competitive in the world market. We need an artistic culture…we also need taste. With taste, we will be able to produce goods and services of far greater value.
>
> (Kong and Yeoh, 2003: 179)

There are complex relationships between the local and the global as mediated through the cultural practices of globalizing cities. The local can become the global. Take the case of what were once local festivals such as Mardi Gras in New Orleans and Carnival in Rio de Janeiro which have moved from local celebrations to mark the end of Lent to globally recognized festivities, part of the world circuit that attracts people from all over the world. The global also comes to the local in the form of traveling exhibits, festivals and sports competitions.

Culture is not only managed in global cities, it is also produced and reproduced. There are cities of global cultural production. Los Angeles is a relatively minor economic player in the city network of producer services and banking, but in the realm of cultural production it is a prime node in an international film industry. The rise of Hollywood is associated with the emergence of a global film industry. In the first third of the twentieth century there were important national film industries in many countries. Now Hollywood dominates a truly global film distribution network. The changes have involved a globalization of American film themes and an Americanization of the international film industry. There are other important cities. Bollywood in Bombay is the center of the national Indian film industry whose movies are widely distributed around the world to the Indian Diaspora. There is also the cost-driven globalization of "Hollywood" movies with the emergence of smaller off-shore production centers such as Sydney, Australia with the necessary technical infrastructure but cheaper costs. The film *The Matrix* was shot entirely in Sydney. While this shift in cultural production is part of the globalization of cultural practices, local cultural industries or practices can also become important elements in global cultures; the experience of Liverpool and Seattle show how music in specific cities can become part of a more global scene. The Beatles and Kurt Cobain moved from being local celebrities to global rock stars as the Merseyside Beat and grunge rock moved out from local clubs to a wider world recognition.

The globalizing cities are important networks in the creation and reproduction of global cultures. The city sites of cultural reproduction (the concert halls, art galleries, sports stadiums, rock concert venues and film festivals) bring global cultures to local places and allow local places to play a part in the creation of global cultures. Table 7.3, for example, lists cities that in the period 2002 to 2003 held film events that had the words *International Film Festival* in their official designation.

Global identities, hybridities of various cultural forms, are more easily produced, maintained and reproduced in global cities where there is a wider range of goods and services from other parts of the world. We could make a distinction between world cities, where global business is conducted and global cities, where both global businesses are undertaken and global culture is more readily available. There are world cities that are more parochial and chauvinistic, such as Tokyo, that lack the cosmopolitanism of truly global cities. The availability of foreign films, the opportunity to sample different cuisines, the overall sense that the rest of the world is available in various forms all mark a truly cosmopolitan city.

Table 7.3 Selected international film festivals, 2002–2003

January	Bangkok, Trieste, Rotterdam
February	Berlin
March	Hong Kong
April	Minneapolis-St Paul, Singapore, Buenos Aires, Istanbul, Jeonju, San Francisco
May	Cannes, Seattle
June	Atlanta, Sydney
July	Okinawa, Melbourne
August	Edinburgh, Sarajevo, Montreal
September	Toronto, Venice, Helsinki, Vancouver, New York, Calgary
October	Cork, Denver, Bogota, Chicago, Valladolid
November	Tokyo, London, Stockholm, St Louis

The consumption requirements of cosmopolitan cities help generate cultural difference. The availability of film festivals and the existence of film audiences ready to watch subtitled or dubbed movies, for example, creates the demand for "local" films that in a sense aids their creation. Movies produced in Bollywood, such as *Monsoon Wedding*, that use Indian actors and technicians, are specifically targeted at non-Indian filmgoers in cosmopolitan cities outside of India. A persistent theme of the movie is the connection between an extended Indian family that spans India, USA and the Middle East.

The cultural practices of global cities can raise the question of authenticity. A sample of so-called sushi restaurants in Washington DC undertaken by the *Washington Post* in 2003 revealed that most of them were Korean rather than Japanese. While entertaining a visitor from Thailand I took her to a range of Thai restaurants in Syracuse. We found it difficult to find anyone in the three supposedly "Thai" restaurants who could speak Thai. Cosmopolitan cities create difference but whether it is authentic difference is another matter. We live in a world of manufactured difference.

Global cities are cosmopolitan, and most globalizing cities seek to become cosmopolitan. The term has an elastic definition. At a basic level it has the sense of the local availability of the global in such forms as international art, foreign film and global fashion. Globalizing cities boast of their range of ethnic cuisines, ethnic festivals and racial mix. The cosmopolitan city leans towards the spectacular. The dazzling opening ceremonies of a major international sporting competition for example, such as the opening ceremonies of the Summer Olympics, provide the platform for global recognition and urban reimagining.

Even terrorism has a global city spectacularization as witnessed in the destruction of the twin towers of the World Trade Center on September 11, 2001.

There is no one-to-one relationship between cosmopolitanism and globalization. We can identify degrees of cosmopolitanism. New York and London are more cosmopolitan than Tokyo. We can imagine a continuum of cosmopolitanism, with certain oil-rich cities of the Middle East at one end where foreign workers live in compounds to minimize their cultural impact on rigidly theocratic societies. At the other end are cities where cultural difference is celebrated and promoted.

Cosmopolitanism is sustained in cities by flows of people and ideas from the outside world. Goods, services and capital can flow around the world without necessarily creating a sense of cosmopolitanism. Economic globalization does not necessarily lead to cultural globalization. However, cultural globalization both creates and embodies cosmopolitanism since local and national culture are directly affected by flows of people, images and ideas.

The presence of foreign people in a city is a good though not perfect indicator of cosmopolitanism. The greater the number and range of immigrants to a city the greater the possibility of cultural mixing. Table 7.4 draws

Table 7.4 The foreign-born in selected cities

City	% Foreign-born	Country	% Foreign-born in country
Miami	50.9	USA	12.4
Toronto	41.5	Canada	18.9
Los Angeles	36.2	USA	12.4
New York	33.7	USA	12.4
Riyadh	31.2	Saudia Arabia	25.8
Sydney	30.8	Australia	24.6
Amsterdam	28.6	Netherlands	9.9
London	21.7	UK	6.8
Paris	12.9	France	10.6
Barcelona	5.6	Spain	3.2
Tokyo	2.4	Japan	1.3
Seoul	0.14	South Korea	1.3

on the work of Lisa Benton-Short, Samantha Friedman and Marie Price who have collected data on the percentage of foreign-born in global and globalizing cities. Notice how Tokyo (and Seoul) have very low percentages of foreign-born compared to London and New York. A high percentage of foreign-born need not necessarily lead to cosmopolitanism. Riyadh in Saudi Arabia has almost a third of its population foreign-born, but the foreigners are demarcated into compounds to minimize their impact on the traditional society. This form of encounter reinforces the forces of reaction and generates the invention of tradition: one response to the cultural other is to segregate them and to stress the pure identity of the non-other.

A cosmopolitan city is not just a variety of different people. Riyadh may have many foreign-born residents but it is not a cosmopolitan city. The cosmopolitan city is a city in which the diversity is not managed to have minimum impact but celebrated for maximum effect. Non-cosmopolitan cities resist diversity while more cosmopolitan cities embrace diversity.

Table 7.4 also notes the percentage of foreign-born in the city as compared to those in the country in which it is located. In all cases, apart from Seoul, the city foreign-born is a much higher percentage of the national foreign-born. Cities are the backbone of cultural globalization and the most important nodes in the networks of cosmopolitanism. But even in the most cosmopolitan cities there is always a tension with national identity. Even in societies such as Australia where there are major cities with substantial immigrant populations, there is a possibility of nativist sentiment becoming a powerful part of the political discourse. The global and globalizing cities are an important setting, as they are the main site for immigrants and the practice of multiculturalism: for some something to celebrate, while for others something to rail against. New populisms and nationalisms compete with liberal discourses for popular appeal and political control. For the former, globalization poses a "mortal threat to their special identity"

(Rupert, 2000). The cosmopolitan city has a complex relationship with the national society, wrapped as it is in global links of capital flows, diasporic communities and cultural connections. National identity is about bounded space, differences and the identification of the other; cosmopolitanism is concerned with unbounded space, similarities and shared global connections. The truly cosmopolitan city is tied as much into the space of global flows as national networks, more part of a global society than a national order.

REIMAGINING THE GLOBAL CITY

The globalizing city is not simply a site of economic transactions, it is a place of global imaginings. Globalizing cities are as much acts of imagination as they are places on the map; they occupy a discursive as well as a geographic space.

In a series of papers I have analyzed the role of place-marketing in globalizing cities (Short, 1999; Short and Kim, 1998; Short *et al.*, 1993). I suggested that regimes of urban representation can be identified that refer to the ideas, concepts and practices that contextualize how and why cities are represented. A dominant regime is urban boosterism; a discursive emphasis on selling the city to a wider business community. However, a specific form of globalizing boosterism can now be identified that is concerned with positioning the city in a global flow of urban images and discursive practices. In a world of hyper-mobile capital and global competition between cities, the reimagining of cities is an active process that represents the city to both internal and external audiences.

There are the wannabe global cities that actively promote a global discourse through specific slogans and tailored campaigns. In recent years the city of Atlanta has used *Claiming Its International Destiny* and *World's Next Great International City*. The history of urban boosterism in Atlanta since the nineteenth century would show a common trajectory, from attracting manufacturing jobs, to luring corporate offices, to promoting global command functions. The 1996 Olympics was the culmination of a long campaign to gain international recognition for the city. On the other side of the global shift, in Shanghai in China there has been an active place promotion with an emphasis on letting foreign investors know that this was a city open to business, as well as highlighting policies on land-leasing and one-stop investment advice (Wu, 2000c and d). There was also a concern to show the city as a cultured, cosmopolitan place. New symbolic urban landscapes such as the Shanghai World Financial Center and the East Concert Hall were built to give a message of openness, culture and global status (Wu, 2000d).

Further down the global urban hierarchy cities adopted a variety of campaigns. For industrial cities in the developed world a major theme of place promotion has been what I have termed "Look No More Factories." City makeovers of industrial cities such as Manchester, Pittsburgh and Syracuse stress the new not the old, the postmodern rather than the merely modern, postindustrial not industrial, consumption rather than production, spectacle and fun rather than work and pollution.

Around the world, recurring themes for those cities attracting investment are pro-business, a well-trained and pliant workforce, good infrastructure, locational advantages and good quality of life for the workers, managers and citizens. Specific campaigns are tailored to the specific niche in the global urban hierarchy. For those seeking to attract more than just routine factory jobs, there is an emphasis on culture and sophistication. Images and slogans are devised that play up the notion of fun city, green city, cultured city and pluralist cosmopolitan city. Tourist cities in contrast, such as Prague concentrate on selling the old rather than the new, where a continuity rather than a rupture with the past is stressed and packaged.

Whatever, the specific campaign chosen, the representation of the city is an active process that affects internal policy and debates as well as external perceptions. Projects that fit in with the new discourse are green-lighted while those that do not are marginalized to irrelevancy. The reimagining of the city involves a subtle reorientation of power within the city. Industrial cities, for example, have a local culture that emphasizes manual work, a collective sense of identity. The choice of postindustrial images to represent the city, on the other hand, undermines the traditional meaning of the city and reallocates status and prestige to different groups in the city. The representation of the city is not devoid of internal, political significance.

The representation of cities also involves a more explicit internal debate within the city since there is always a shadow to the new light that is shone on the city. Things not spoken about, images not presented and discourses not raised are as much a part of urban reimagining as the images and debates actively promoted. Some practices are presented as out of date and irrelevant while some groups, such as manual industrial workers in formerly industrial cities, are seen as redundant. Low-waged workers vital to the smooth functioning of the city, yet unconnected to the new global imaginings, are denied discursive space in a parallel to their spatial and social marginalization in the social geography of the city. In this shadow representation there is an apportionment of blame to certain groups as well as the issue of social control and the foreclosure of alternatives.

The use of natural analogies, such as revitalizing the heart of the city, not only dramatizes the endeavor but naturalizes it so that to argue against it would be unreasonable, beyond the reach of common sense. The silences and languages of urban representation are as supercharged with messages of social control and political power as the promotion campaigns of globalizing cities. Urban representations and their silences are important texts in the rewriting and reimagining of the city. And like all urban restructurings some people gain and some people lose and it is important to be aware of the redistributional consequences.

8 Going for gold
Globalizing the Olympics, localizing the Games

The modern Summer Olympic Games are an interesting case study of some of the more intriguing connections between the city and globalization. They are, at one and the same time, global spectacles, national campaigns and city enterprises. They are "glocal" events of national significance.

There are many relationships between global cities and the Summer Olympic Games. Bidding and hosting the Olympic Games involves a rewriting and reshaping of the city. The Games provide an important platform for place marketing as cities seek to achieve international recognition and global city status.

To understand the connections between the city and globalization, it is necessary to provide a historical context for the modern Olympics. The modern Olympic Games were inaugurated as a national political project. They were not devised to replace nationalism, but to channel it. The initiator of the modern Games, Baron Pierre de Coubertin (1863–1937), came from a wealthy French family, and his early interest in sport was situated in a particular class and gendered space and deeply embedded in national politics. He saw the 1870 French defeat in the Franco-Prussian conflict as a bitter blow to French self-esteem. Part of the reason for the defeat, he believed, was in large part the poor physical condition of the French troops, especially compared to the better-trained, fitter, German troops. His reorganiz-ation of French sport was an attempt to rebuild the physical fiber of young Frenchmen. He drew upon the sports at English public schools as one of his models for youthful, male athleticism.

His national concerns widened to the promotion of sport as a forum for peaceful internationalism. After reorganizing a number of French national sports associations, he established the Union des Sociétés Françaises de Sports Athlétiques, and in 1892 he revived the notion of the Olympics Games as a forum for international athletic competition. At the end of the nineteenth century, there were few international sporting organizations or events. Sports were organized along local and national lines. His proposal was part of a more general discourse of internationalism.

The early internationalization of sport in the modern era arose from a wider context of both global international conflict and co-operation. From around 1870 to 1914, the period Gollwitzer (1969) refers to as the age of imperialism, there was both a more pronounced nationalism and a keen internationalism. There was growing rivalry between European powers and the USA for overseas markets and global dominance. Economic and political rivalry reached its peak with the European partition of Africa and the US annexation of the Philippines, Puerto Rico and Cuba. The major powers were colliding in global space. However, there was also growing co-operation between countries as their increasing interaction in global space also promoted shared projects of political and spatial management. Economic integration was an important force. It was a time of low tariffs, an international labor market and relatively free capital mobility. There was a wave of globalization as space-time collapsed with the advent of the railways, the telegraph and internationally organized postal unions. The period is marked by an internationalism, an era studded with international meetings, conferences, conversations and conventions that laid down international standards and arrangements. Four brief examples: First, at the Congress of Bern in 1874, twenty-two countries signed the International Postal Convention. That event set the stage for regular and cheap internal postal services. Second, the International Workingmen's Association (the first incarnation 1864–76, the second 1889–1919) was formed to give substance to international solidarity of the fledgling workers' movements. Third, the Hague Conventions of 1899 and 1907 codified the international rules of war. And finally, in sports, the Federation of International Football Associations (FIFA) was formed in Zurich in 1904.

The resurgence of the Games was part of this wider and broader trend of internationalism. Coubertin believed that reviving the ancient Games would counter the worst excesses of nationalism and provide a peaceful forum for international competition.

GOING GLOBAL

The global reach of the early Olympics was limited. At the beginning of the twentieth century sports did not enjoy the exalted position it now holds. A key component of both globalization and nationalism has been the general diffusion of nationally organized sports throughout the late twentieth century world. Sport has become an international phenomenon, a cultural form that transcends national language while also expressing national identity. The Olympics Games have done much to stimulate national sports organizations and especially participation in sports events represented in the Games. Sport, like English, has become a global language and globalization is embodied in such international sport practices as the Olympic Games. But in 1896 this was all in the future.

An early limiting factor to the global diffusion of the Olympic Games was the cost and difficulty of international travel. Athletes had to pay their own way and the cost and time taken were often prohibitive. The 1904 Games held in St Louis were considered so far away that only twelve countries were involved and, of the 689 athletes, 432 came from the United States. The 1904 Games were for all intents and purposes, the US national games. In 1912, it took the US team ten days to get to Games in Stockholm, Sweden and, in order to keep training during the long sea voyage, they built a 100-yard, cork running track and made a canvas swimming pool on board the ship. The Games broadened only as international travel become easier, quicker and less expensive. In 1932 the economic depression limited the participation in Los Angeles to fewer than half the athletes that had competed in Amsterdam in 1928. The costs of international travel were so prohibitive that the Japanese government offered to subsidize the travel of teams intending to visit the proposed 1940 Olympics in Tokyo.

The Coubertin international project was restricted to only a few countries, essentially the richer countries of Europe and the USA. At the 1900 Games in Paris, only twenty-six countries competed. Throughout the twentieth century there has been a steady increase in the number of countries competing (see Table 8.1). The figure increased to fifty-nine by the time of the London Games in 1948, and in Sydney in 2000 there were athletes from 199 countries. By the time of the Sydney Olympics, the Games had become truly global, and most countries of the world competed.

Table 8.1 The Olympic Games

Date	Host city	Participants	(women)	Countries	Sports	Events
1896	Athens	200	(0)	14	9	43
1900	Paris	1,205	(19)	26	24	166
1904	St Louis	681	(8)	13	16	104
1908	London	2,035	(36)	22	21	110
1912	Stockholm	2,547	(57)	28	13	102
1920	Antwerp	2,668	(77)	29	21	154
1924	Paris	3,092	(136)	44	17	126
1928	Amsterdam	3,014	(290)	46	14	109
1932	Los Angeles	1,408	(127)	37	14	117
1936	Berlin	4,066	(328)	49	19	129
1948	London	4,099	(385)	59	17	136
1952	Helsinki	4,925	(518)	69	17	149
1956	Melbourne	3,184	(371)	67	17	151
1960	Rome	5,346	(610)	83	17	150
1964	Tokyo	5,140	(683)	93	19	163
1968	Mexico City	5,530	(781)	112	18	172
1972	Munich	7,123	(1,058)	121	21	195
1976	Montreal	6,028	(1,247)	92	21	198
1980	Moscow	5,217	(1,124)	80	21	204
1984	Los Angeles	5,330	(1,567)	140	21	221
1988	Seoul	8,465	(2,186)	159	23	237
1992	Barcelona	9,634	(2,707)	169	24	257
1996	Atlanta	10,310	(3,513)	197	26	271
2000	Sydney	10,651	(4,069)	199	28	300

While the number of countries participating in the Games has increased, the siting of the Games still reflects the Euro-US bias. Only seventeen countries will have held the Games from 1896 to 2004, five of them twice: Australia, France, Germany, Greece and UK. It was only in 1956 that the first Games were held outside of Europe and the United States. It was only in 1964 that the Games were held in Asia, and of the twenty-five Games, up to Athens 2004, only four have been held outside Europe and North America. Beijing in 2008 will be only the third time that the Games will be held in Asia. Hosting the Games means either existing substantial or proposed infrastructural investments that only few countries in the world can afford or are willing to undertake.

The partial nature of the early Olympics was also due to differences in wealth. Athletes for the early Games had to pay their own way. Few could afford such expense. And even when the costs were borne by national committees, effective participation was limited. At the first Olympics, there were athletes from only fourteen countries. While the Games have broadened in participation, it is an unequal participation in two regards. First, some countries send more athletes than others. Sending athletes to the Olympic Games is an expensive undertaking and richer countries tend to send more than poorer countries. Second, while many countries compete, fewer succeed. Although there are exceptions, the richer countries still do better than the poorer countries. Richer countries can not only send more athletes, they can also afford the necessary expenditure in sports development and training that ensures success. Success at the Games reflects the profound global inequalities in wealth. Of the 927 medals won at the Sydney Games in 2000, just five countries won 357: USA (96), Russia (88), China (59), Australia (58) and Germany (56). The nationality of medal winners reflects global differences in wealth: Europe and North America won the most medals, 463, or approximately 50 percent. Sub-Saharan African countries, in contrast, won only twenty medals, or just over 2 percent.

The fact that the USA, Russia and China win most of the medals is in part a function of population size. If we standardize for population size then a very different picture emerges. Table 8.2 lists the top ten countries ranked in terms of all medals per million population won at the Sydney Games. The data picks out very small countries that won just one medal, such as Barbados and Iceland, as well as those small countries that picked up a lot of medals such as Australia and Cuba. In both of these countries there was a national commitment to sporting excellence and Olympic achievement. Australia also gained from home advantage. Given the relative poverty of Cuba, its medal tally is truly astonishing. The case of Cuba highlights the exception to the general rule that national wealth is a way to Olympic success. Countries either have to be wealthy to succeed and/or they need to devote significant resources to selected Olympic sports. The UK, for example, which won only fifteen medals in Atlanta, won twenty-eight medals in Sydney. In between these two Games, the British government spent almost $1.5 billion on Olympic athletes.

Table 8.2 2000 Olympic Games: medals by national population

Rank	Country	Total medals	Population
1	Bahamas	2	294,982
2	Barbados	1	274,059
3	Iceland	1	276,365
4	Australia	58	19,164,620
5	Jamaica	7	2,652,689
6	Cuba	29	11,141,997
7	Norway	10	4,481,162
8	Estonia	3	1,431,471
9	Trinidad & Tobago	2	1,175,523
10	Hungary	17	10,138,844

Success at the Olympic Games reflects wealth and national spending on sports. Countries with few resources and little spending are less successful. And even when athletes from poor countries achieve success it often comes after using the training and coaching facilities in the richer parts of the world. Even as participation in the Games becomes more global, success at the Games becomes more uneven. In effect, the Games reinforce the unequal distribution of resources in the world by the unequal participation of different countries and their unequal success in mounting the medal podium.

DEMOCRATIZING THE GAMES

The Olympic Games, both in terms of organization and participation, were organized along elitist, masculinist principles. In order to minimize nationalism in the running and organization of the Olympic Games, Coubertin established an International Olympic Committee (IOC) chosen by invitation from a small pool of rich, white men. It was consciously insulated from any electoral process. This Committee was and remains the principal governing body of the Olympic Games. The early members were either rich, titled or well-connected: preferably all three. Members were expected to pay their own travel expenses and make a financial contribution to running the IOC. In 1896 the IOC consisted of fifteen men from only thirteen countries: Argentina, Belgium, Bohemia, France, Germany, Great Britain, Greece, Hungary, Italy, New Zealand, Russia, Sweden and USA. The members included a Duke, a Lord, a Baron, a Count, a General and a Major.

Over the years the IOC membership increased but scarcely broadened. By the time of the 1948 Games in London the IOC had increased to sixty-six men from forty-one countries. The membership had widened but only insofar as it now included a Duke, a Marquis, two Lords, two Barons, two Sirs, five Counts, three Princes, two Generals, a Pasha and a Raja. By 2002 the membership had widened to 123 members from eighty-two countries. There was now only a smattering of the titles: a Princess, a Princess Royal, a Dona Infante and a Grand Duke. But still only seven women. Membership is dominated by the rich and the connected. The biographies listed at the IOC website reveal a membership dominated by rich business people: corporate lawyers, presidents of business groups and company directors figure largely (IOC, 2002). The IOC is not a representative body: it is a rich man's club in which personal and business connections wrap around Olympic business (Jennings, 1996; Jennings and Sambrook, 2000). One example: Juan Antonio Samaranch, from Barcelona, became President of the IOC in 1980. In October 1986 Barcelona was awarded the 1992 Summer Olympic Games. In 1987 he also became president of the largest Catalan financial institution La Caixa and three years later became president of the expanded institution, La Caja de Ahores de Barcelona.

While the discourse of democratization has swept the world, the IOC remains largely insulated from democratic accountability. Along with the IMF, WTO and World Bank, the IOC stands as an important, undemocratic, unaccountable globalizing organization. There has been some minor reform in recent years, prompted by the scandals of corruption discovered in the bid of Salt Lake City, but the IOC remains an elitist organization. Headquartered in Lausanne, Switzerland, it forms an unelected para-state, insulated from reform.

It is important to note that there were forms of international sporting organizations that provided an alternative to the original Olympics. Over 100,000 spectators saw the first Workers' Olympics held in Frankfurt in 1925. These Workers' Games were repeated in 1929, 1933 and 1937. But it was the other, more elitist, patrician-organized games that became the dominant and eventually sole form of Olympic Games.

INTERNATIONAL BUT RESTRICTED

The Games were initially organized along exclusionary principles. They were biased in both class and gender terms. Coubertin drew upon the nineteenth-century, English, Victorian, bourgeois depiction of sport as a gentlemanly pursuit. The amateur code in England was explicitly defined to exclude the "lower classes." The English influence was also apparent in the fact that cricket was a recognized Olympic sport in the very first Olympics in 1896. When codified into the Olympic charter, athletes who were considered professional were banned from competition. The amateur code, in effect, operated against waged workers and lauded the athletic pursuits of the wealthy into a discourse of nobility and moral superiority. The exclusion was particularly effective in early Olympics. When it was discovered that Jim Thorpe, the Decathlon Gold Medal winner in the 1912 Stockholm Games, had been paid small amounts of money for playing baseball while on holiday three years earlier, he was stripped of his medal. It was awarded posthumously in 1983.

The strict amateur code began to break down as Olympic success necessitated full-time training and commitment. It was no accident that the UK, where the amateur cult has persisted longest, has had proportionately less Olympic success than either its population or national wealth would predict. In many countries, the definition of amateur became elastic. The entry of communist countries in 1952 redefined the meaning of amateur. From the 1950s onwards, the strict amateur code was more often breached than honored and for many years, the Games became exercises in hypocrisy (Strenk, 1981). In 1978 a rule change in the Olympic Charter allowed athletes to earn endorsement money if the money was given to the National Olympic Committee or the appropriate International Sports Federation. The issue for the forward thinkers in the IOC became not for or against commodification, but how much could be gained from the process. The ascendancy of Samaranch to the presidency of the IOC (a post he held until 2001) marked an increasing concern with the commercialization of the Olympic Games.

Only men competed in the first Games held in 1896. Drawing on the gendered English public school system, Coubertin saw no room for women in the Olympic Games. Women's participation had to be fought for and their representation has persistently lagged behind men's (see Table 8.1) with many events being closed to women. The evolution of the Games reveals a steady increase in women's participation, but these have been fought for, not given. Coubertin resisted female athletes; only a few female golfers and tennis players were allowed to compete in 1900, and female swimmers were only included in 1912. The refusal of the IOC to accept women's athletic participation at the 1920 Antwerp Games led to a breakaway women's movement. Women's Olympic Games were held in Paris in 1922, seven countries attended and subsequent Women's Games were held in Gothenburg in 1926, Prague in 1930 and London in 1934. In order to get into the Olympics female athletes performed for IOC delegates in 1930. The American delegate noted:

> I personally saw groups of young girls in the scantiest kind of clothing trotting around the fields or running tracks, engaging in 100 metre runs, taking part in the broad jump, and hopping about in all kinds of athletic and gymnastic movements, and to my direct statement as to whether or not such character of exercise was not bad for them, the answer was that on the contrary, it was good for them.
>
> (Quoted in Guttman, 1992: 49)

Women's participation had to fight against such attitudes and even as late as the 1984 Games in Los Angeles, women's participation was only 22 percent of men's. By the 2000 Games this figure had increased to 38 percent.

Race has played a complicated role in the Olympics. The elitist nature of the early competitions would lend some credence to Cannadine's assertion that class as much as, if not more than, race and ethnicity was a key factor in the social distinctions in the imperial powers (Cannadine, 2001). But, while there was never any formal racial exclusion by the Olympics, the Games have often been racialized. At the 1912 Games the British press complained about the Negro and Indian athletes competing for the USA. A Nazi commentator in the 1930s complained that modern sport was "'infested' with Frenchmen, Belgians, Pollacks, and Jew-Niggers" (quoted in Guttman, 1992: 54).

The infamous Berlin Games of 1936 espoused a philosophy of Aryan superiority. A boycott campaign was organized against German treatment of Jewish athletes in Canada, France, UK and USA (Bachrach, 2000; Mandell, 1971). Despite the protests and obvious instances of Nazi repression and exclusion of Jewish athletes, the IOC continued to support holding the Games in Berlin. Coubertin gave a radio message in 1935 declaring his confidence in the Games and, in return, the Nazi regime nominated him for a Nobel Peace prize. Pamphlets were written at the time promoting the boycott; but the IOC decided to go ahead with the Berlin Olympics.

There was never an IOC policy of exclusion by skin color. However, IOC policy towards South Africa reveals an interesting picture. South Africa was a member of the Olympic Movement but also practiced apartheid, which went against the Olympic code of equal access. The less than convincing defence of the South African National Olympic Committee (SANOC) in the early 1960s, that all the best athletes just happened to be white, was a clear violation of Olympic principles. The IOC voted to suspend SANOC in June 1964 and South Africa was not allowed to send any athletes to the 1964 Games in Tokyo. The South Africans lobbied for a return and in 1968 the IOC dropped the suspension. The African countries promised to boycott the 1968 Games, a boycott that was supported by the USSR and many Eastern Bloc countries. Consequently, the IOC quickly changed its votes and SANOC's suspension became an expulsion by a narrow margin of 35 to 29. At Montreal in 1976, athletes from twenty-eight African countries were recalled because New Zealand, a participant in the Games, had played rugby against South Africa.

THE GAMES AS GLOBAL SPECTACLE

The first Olympic Games of the modern era took place in Athens in 1896 where an enthusiastic and partisan crowd saw 200 participants compete, all men, from just fourteen countries. It was an important national event, but it had limited international impact. It took a long while for the Games to become global spectacles and the process is intertwined with development of mass media, particularly television.

The press always reported the Games, but the coverage was initially slight. In 1928 the UK *Daily Express* noted, "The British nation, profoundly interested in sports, is intensely uninterested in the Olympics" (quoted in Guttman, 1992: 48). It is impossible to think of a British or indeed almost any national newspaper repeating this assertion today.

Media coverage increased by the 1920s as the Games grew in scope. There were 700 news reporters at the LA Games in 1932. The Nazis, always intensely aware of the power and significance of spectacle, gave the 1936 Games the first multimedia coverage. There was television transmission to twenty-five TV halls in Berlin and short-wave radio broadcasts to forty counties. An official movie was made. The in-house Nazi cinematographer, the immensely gifted Leni Riefenstahl, paid homage to the physicality of the event in the film *Olympia* that was released in 1938.

There were no cameras at Melbourne in 1956 because the TV companies refused to pay for what they regarded as public property. Avery Brundage, at that time the IOC President, remarked, "the IOC has managed without TV for sixty years and believe me – we are going to manage for another sixty." The remark did not prove to be prescient. After Melbourne, television coverage became an integral element of the Games.

The growth of the Games and their increasing globalization was connected closely to television coverage that could transmit the images worldwide. For the 1960 Rome Games, CBS paid $660,000 for the right to fly film from Rome to New York, while Eurovision transmitted the first live coverage of the Games. The Italian Olympic Committee earned $1.2 million from the deal. There has been a steady increase in the coverage ever since. Only twenty-one countries saw television coverage of the Rome Games, but by Atlanta in 1996, this had increased to 214. Few countries in the world are unable to see the Summer Olympics. More than 3.7 billion people watched the Sydney Olympics from 220 countries. The typical viewer watches the Games eleven times, resulting in a combined viewing audience estimated at 36 billion.

Selling television rights has become an increasingly important part of funding the Games. In Munich in 1972 less than 10 percent of the revenue of the Games came from television companies, but by Atlanta in 1996 this had increased to almost 40 percent. The absolute amounts have grown on average 30 percent each Olympiad, from $40 million in 1972 to $556 million in 1996. In a package deal, the NBC paid $3.5 billion to cover the Sydney Olympics, Athens, and Beijing as well as the Winter Games of 2002 and 2006. Television revenues currently provide 55 percent of all IOC's marketing revenue. US TV companies, in particular, account for 60 percent of the total worldwide rights. The Summer Games are now thoroughly corporatized, providing a huge global audience of consumers and a global opportunity to sell goods and services around the world.

The notion of the Games as global spectacle needs to be treated with some care since the Games are seen differently in different parts of the world. Apart from the opening and closing ceremonies most national audiences see different Games (Tomlinson, 1996). To be in the USA or Hungary and see the Games is to see very different Games. You can see hours of fencing in Hungary, for example, while this scarcely merits much attention in the USA. People in different countries quite literally see different Olympics as their coverage concentrates on their national teams and representatives.

Increasingly, the television producers have crafted individual stories of specific athletes to add human interest and wider appeal, resulting in longer average viewing. Within this general framework, the coverage of the Games varies around in the world. Likely British medallists, for example, are given extensive coverage in the UK, while the US television coverage is both national and market-driven. Potential US medallists are likely to attract a wider TV audience, but so are sports "personalities" from other countries. Uplifting tales of redemption, narratives of recovering from adversity and tales of human drama may also receive full attention. US television coverage is not only nationally biased, but also market-biased. Thus, women's gymnastics receives more coverage than boxing in the USA even when the national boxing teams are doing better than the national gymnastics team, because more people with higher spending patterns watch gymnastics more than boxing. Ever since Olga Korbut's performance in Munich, women's gymnastics have been a firm favorite of US television coverage.

The coverage of the spectacle is also programmed for maximizing audiences. Apart from the opening and closing ceremonies, fewer and fewer events are covered in real time, and it is not simply because of the time differences between different parts of the world. We were reminded of this in the 1996 Atlanta Games when the US women's gymnastics team was successfully competing for a medal in the team event. The competition, conveniently scheduled for prime time Saturday night, seemed to the television audience a long-drawn-out event, lasting over three hours to a medal climax at around 11 p.m. Eastern Standard Time. Even for viewers in the same time zone, the coverage had been delayed and extended to maximize the TV coverage and hence the viewing audience's exposure to adverts. Long after the "real time" event had finished the television "event" was still running its long course with incessant advertisements.

In their detailed study of Olympic television coverage McCollum and McCollum (1981) revealed that over 55 minutes of advertising was shown during five hours of coverage. Only 24 percent of the five hours was devoted to sports, while 61 percent was devoted to network adverts, personal interviews and reruns. They described the coverage as "commercials, network program plugs, and a continual stream of non-sports features and interviews." The Olympic Games are a global spectacle perhaps, but a nationally biased, commercially driven, global spectacle.

THE CORPORATIZATION OF THE GAMES

There is both an implicit and explicit corporatization of the Olympics. Implicitly, it occurs through the interlinked directorships and connections of the IOC members. Developers and financiers rather than minor royalty now make up the IOC and their business interests mesh seamlessly with their Olympics position (see Jennings, 1996). The more explicit role of corporations in the funding and direction of the Games is often dated to the 1984

LA Olympics, but there is an even longer connection between the Games and commerce. The very first modern Games in 1896 relied upon a businessman, George Averoff, to finance the refurbishment of the Olympic stadium in Athens. In 1912 ten Swedish companies were given sole rights to the photographs of the Stockholm Games. The Rome Olympics had forty-six private sponsors, and there were 168 official products at the Montreal Olympics. And even the 1980 Moscow Olympics endorsed 200 products. However, the role of corporate sponsorship changed dramatically in the 1984 Los Angeles Games. In order to defray the costs the Los Angeles Olympic Committee (LAOC) signed up thirty-four corporate sponsors including Coca-Cola, Mars and Anheuser Busch. Each paid between $4 million and $15 million for the exclusive right to market their products with the Olympic logo. The LA Olympics turned a modest profit. It was not that sponsorship was new, but that it was more dominant after 1984. Prior to 1984 sponsorship never accounted for more than 10 percent of total revenue. Los Angeles was the first truly corporate Games: 20 percent of total revenues of $1124 million came from corporate sponsors. It marked the beginning of an upward trend in corporate funding. In Atlanta, of the combined revenues of $1686 million, almost 30 percent came from corporate sponsorship.

The cost of the Games has grown enormously. In 1960 Rome spent $50 million on public works, the Munich Olympics in 1972 cost $850 million, the 1976 Games in Montreal cost $1.5 billion and the cost of the 1980 Moscow Games ranges in estimate from $2 billion to $9 billion. Ticket sales and sales of the coins, stamps and official mascots – the traditional sources of revenue – failed to cover the spiralling costs. Corporate sponsorship along with television revenue provided the necessary revenue to host increasingly lavish Olympics.

The LA experiment was so successful that it became a model for a marketing program introduced by the IOC. The Olympic Program (TOP) drew up long-term agreements between the IOC and large corporations from different business sectors. The TOP sponsors for the period 2001 to 2004 are SchlumbergSema, McDonald's, Coca-Cola, Panavision, Visa, Xerox, Time, Kodak and John Hancock. Under the TOP agreements, the IOC receives long term financing while the sponsors get access to a worldwide audience in an association with the Olympics, one of the most recognized and positively perceived "brands" on the planet.

The corporatization has involved a narrowing of the business interests to a few giant, global corporations. In Montreal 742 enterprises advertised with the Olympic Games, by Sydney in 2000 this had fallen to less than 100. There has been a narrowing and deepening of corporate sponsorship. For the top corporate sponsors, the global coverage of the Olympics Games provides the opportunity to spread global recognition and appreciation of their products and services. The Games have become an important vehicle of economic globalization, a platform for the penetration of selected corporations into global markets and global consciousness.

The increasing importance of major corporations to the IOC has affected IOC policies. Their quick, although weak, response to the scandals emanating from the Salt Lake City corruption charges was in large part due to their need to reassure corporate sponsors. The Olympics is a brand with a high value because of its positive associations: corruption charges undermined the value of the brand. The corporate sponsorship has also influenced the siting of the Games. The major corporations have been very eager to get the Games into China as a strategy of promoting their products and name recognition to one of the largest faster-growing markets in the world. Beijing came very close to getting the 2000 Games, which went to Sydney. From the corporate standpoint there are only 17 million consumers in Australia but over a billion in China. It came as a relief, but not much of a surprise, to the sponsors that Beijing was successful in landing the 2008 Games.

THE INVENTION OF TRADITION

The IOC and major media present the Olympics as a "pure" global event above national politics and grubby commerce. The dominant narrative is of an event born in purity and full of innocence. National politics and commercial considerations are often contrasted to the mythical narrative of the sacred quality of the Olympic tradition, and the pure Olympic ideal. This attempted distancing of the Olympics from the mundane world of politics and money is fictitious but nevertheless an important Olympic myth. There never was any innocence (see Espy, 1979; Hargreaves, 1986; Hill, 1992; Tomlinson and Whannel, 1984). But this still leaves the question of why and how this discourse persists. It exists in part because we want it to be true. In a rapidly changing world where commerce and politics produce a necessary pragmatism it is important to have an unchanging reference point. The more things are changing the more important it is to have a fixed perspective. A once pure Games is a case in point. In part, it also exists because it is a message reinforced by those selling the Games at the global, national and local levels. As the world becomes more politicized and more commercial the more the "pure" Games becomes an important commercial product.

SITING THE GAMES IN NATIONAL DISCOURSES

The internationalization of sports and the creation of a global community is embodied in the growing importance of the Olympics Games. However, the Games also reinforce nationalism: they exaggerate as well as channel nationalism.

The principal organizing framework of the Games is national. In the very first modern Games people could enter as individuals, but since 1912, each competitor has to be part of a national team. Only athletes chosen by national Olympic organizations can compete. National representation has been the dominant semiotics of the Games: athletes represent specific countries, they must wear national uniforms; in the opening ceremony athletes enter as members of a national team, and medal winners are rewarded with the raising of their national flags and the playing of their national anthems. The Games are a ritualized nationalism made vivid when winners run round the stadium holding aloft their national flag. The Games intensify national feelings rather than transcend them.

THE NATION-STATE AND THE GAMES

The Games provide a bridge between the nation-state and the international community. Hosting the Games has become an often used and well-recognized form of redemption. There are a number of such cases where the Games have been hosted by countries either previously excluded or marginalized. After the First World War, Germany was not invited to participate in the Antwerp games of 1920 nor the Paris Games of 1924. In 1931 the IOC awarded the 1936 Games to Germany in large part due to the influence and lobbying of two influential German sports leaders, Theodor Lewald and Carl Diem, who were eager to relocate Germany in the international community. Subsequent political upheavals in Germany resulted in the Nazis being in charge of the Olympics of 1936. Germany was again successful in re-establishing international credibility when it was awarded the 1972 Games. Other former Axis powers have also cemented their post-war international citizenship; the Rome and Tokyo Games of 1960 and 1964 involved the rehabilitation of national reputations sullied by their Axis involvement in the Second World War.

Hosting the Games is also a platform for the internationalization of the nation-state. The Seoul Games of 1988 were explicitly used by the South Korean government to internationalize the society as well as to win a stamp of international approval for an emergent power. And the 2001 awarding of the 2008 Games to Beijing marks, as well as China's entry into the World Trade Organization, the culmination of China's growing international involvement and global citizenship.

As the Games became more globally important, success in the Games became a recognizable element of national prestige. This was heightened during the Cold War, when not only competing countries were involved but also diametrically opposed political ideologies. From 1952 to 1988 the Olympics Games were an important arena for Cold War politics. The Helsinki Games of 1952 marked the entry of the Soviet Union into the Olympic movement. From then until the Seoul Games of 1988, success in the Olympics was part of a broader ideological struggle. The USA and USSR competed to win the most medals as a form of political validation. It was not only the two superpowers that were involved. From 1972 until 1988, when East and West Germany were represented by two different teams, the East Germans in particular spent considerable sums and devoted considerable resources to achieving Olympic success. The few remaining Communist

powers, China and Cuba, still devote much attention to Olympic participation and medal success as a form of political substantiation.

The goal of Olympic success was and is not restricted to one side of the
Iron Curtain. Take the case of Australia. It was very successful in the early
post-war Games and in the 1956 Games took a total of thirty-five medals
(thirteen gold) and was placed third in the medal table. Australia defined itself
in large part in terms of international sporting achievement (Magdalinski,
2000). The poor showing of Australia in the 1976 Games, where it won only
five medals (none of them gold) and was placed twenty-fourth in the medal
table, stimulated a national debate that led to the establishment of the Sports
Institutes and considerable public spending on the training of elite athletes
for Olympic success. The investment paid off. In 1996 Australia finished
twelfth in the medal table with twelve medals and in the 2000 Olympics,
where home advantage was important, it was fourth in the medal table with
fifty-eight medals.

Participation and success in the Games is an important form of national
prestige and standing. Success in the Games can become an important
element of national identity. For both Cuba and Australia, small countries
with large Olympic success, Olympic participation and medal success have
become a vital strand in their national identities and national representations. Sports cultures play an important role in national representations.
Olympic success has become an integral part of how countries present
themselves to other nations. National cultures are intimately connected to
sporting achievements.

While participation in the Games is used to promote national prestige,
non-participation is used to make national political statements and score
points in the international community. There were vigorous campaigns to
boycott the 1936 Berlin Games in protest against the treatment of Jews in
Nazi Germany. The boycott and the threat of boycott also became important elements of the post-war Olympics. Egypt, Lebanon and Iraq boycotted
the 1956 Games in protest against the Anglo-Israeli take-over of the Suez
Canal. The same Games were also boycotted by Spain, Switzerland and the
Netherlands to protest the Soviet invasion of Hungary. The threat of boycott by African and Eastern European countries in 1968 forced the expulsion
of South Africa from the Olympic Movement. Nineteen African countries as
well as Guyana and Iraq boycotted the 1976 Games in protest against New
Zealand, which had earlier that year played rugby against South Africa. The
US boycotted the 1980 Moscow Games in protest against the Soviet invasion
of Afghanistan. The boycott included West Germany, China, Japan and
Canada. In retaliation the USSR and Eastern Bloc countries boycotted the
1984 Los Angeles Olympics.

Boycotts mark much of the post-war era of the Games which has been
dominated by the issues of the Cold War and South Africa. In recent years,
boycotts have declined. The main points of friction have dispersed. The
Cold War has disappeared as a central political fracture while the Olympic

racial divide has been softened by South Africa's move into a post-apartheid state.

THE NATION, THE STATE AND THE GAMES

The relationship between the nation-state and the Olympic Games is complex because of the different levels of the state. In the Atlanta Games, for example, the US government had less of a role than the city of Atlanta and the state of Georgia. In contrast, the 1988 Seoul Games featured the central state as a major player in the bidding for and hosting of the Games. The Munich Games of 1972 involved a concerted effort by the three levels of government, the federal government, regional government and the city, which together bore the cost of the Games. In other cases, an uneasy compromise exists between the different levels. In the Barcelona Games of 1992, for example, the Catalan government was eager to promote the Games as a Catalan event while the Spanish government wanted some recognition that the event was taking place in Spain. In the Sydney Olympics of 2000, the New South Wales government was the most important state unit promoting and underwriting the Games. The Australian federal and Sydney city governments had little say or influence. Different levels of the nation-state are involved in the Summer Olympics depending on the political culture, federal arrangements and administrative structures of the host country and host city.

We can make a distinction between the state and the nation. The state is a formal political jurisdiction, while the nation is a group identity pertaining to specific territory. States may contain more than one nation. Dominant nations tend to monopolize nation-state representations, including when they host the Olympic Games. Throughout most of the history of the modern Games, states' minority nations were given little space to be represented. The 1936 Berlin Games, for example, presented the picture of an Aryan nation. However, in recent years, hosting the Games has allowed opportunities for different nations of a state to be represented. In the 1992 Games in Barcelona, for example, Catalan nationalism was an important part of the rhetoric of the Games. The Catalan national anthem was played at the opening ceremony, and both a Catalan and Spanish nationalism were employed in the presentation of the Games. In a mutually beneficial compromise between different parts of the nation-state, Catalan sensitivities were embodied into the Games while Spain was presented as a modern, liberal, pluralist democracy (see Hargreaves, 2000). The success of the Barcelona Games was in part the "win-win-win" of the three scales of nation, state and city. Spain enhanced its reputation as an efficient democracy, Catalonia got an economic boost and a sense of identity in the wider world and Barcelona received an urban makeover and improvements in its infrastructure and global connections.

The 2000 Olympics featured national representations of indigenous Australians. The opening ceremonies privileged aboriginal cultural practices. The athletic success of Kathy Freeman provided an important peg on which to hang a number of narratives about the (changing) role of indigenous peoples in Australian society (Hanna, 1999). Compare this with the 1956 Olympics in Melbourne where there was but a single representation of Aboriginal Australia in an apparently otherwise all-Anglo society. More recent Games provide greater opportunities for wider, alternative national narratives.

The Games also may bind a country together. One of the most important rituals, first started by the Nazis in the 1936 Games, is the torch relay from Greece to the host country, culminating in the eventual entry into the stadium to coincide with the official opening ceremony of the Games. The torch's ritualized journeys throughout the country generate enthusiasm for the Games, prime commercial markets, and create spectacles of national cohesion and unity. As the torch crosses the country, it binds regional differences in a shared and common experience.

SITING THE GAMES IN A CITY CONTEXT

The Summer Games are named after and take place in specific cities. Originally, Coubertin envisaged the Games for capital cities only. The capital-city fixation remains; of the twenty-four host cities from 1896 until 2000, fourteen of them have been capital cities. Out of the first seven games, only two took place outside a capital city.

The very first Games of the modern era were sited in Athens in order to forge the link with the idealized Games of the ancient world. The city did not welcome the Games. A wealthy Greek businessman reluctantly put up the money to restore the stadium. Soon, however, the early modern Games became an integral part of city boosterism and city image-making. The second Games, held in Paris in 1900, were associated with the Exposition Universelle, a world's fair. The Olympic Games became the city spectaculars of the twentieth century, carrying on the tradition of the world's fairs that, beginning with the Great Exhibition of 1851 in London, were the city spectaculars of the nineteenth century. There is a connection between the two spectacles. Coubertin visited Chicago's Columbian Exhibition in 1893 as an observer for the French Ministry of Education and drew inspiration for hosting international events through his impressions. Two of the early Games overlapped in specific cities with world's fairs. The French government agreed to hold the 1900 Games in association with the planned world's fair. Their spatial arrangement revealed the relative prestige of the two events: the Exposition Universelle was in the center of the city while the Olympic events were on the periphery. A few years later, the boosters of St Louis organized a world's fair with the new Games thrown in for added international appeal. The 1904 Summer Games was an extra attraction to the St Louis World's Fair.

With the first bid of the City of Los Angeles, the bidding to host the Games became more separated from national political considerations and included the hopes and aspirations of particular city regions. The city made a bid for the Games as early as 1923 at the instigation of an IOC member from Los Angeles, William May Garland. He was chairman of the local Community Development Association and worked closely with the city's business and political elite, especially Harry Chandler, owner of the *LA Times*. Garland was also president of the Chamber of Commerce. The impetus for the Games came from local boosters, business leaders and real estate interests, a constellation of interests referred to as the "urban growth machine" (Jonas and Wilson, 1999; Molotch, 1993). The 1932 Games explicitly were employed to boost the city's image, economy and business fortunes. Proposed at the time of boom, they also were used to stimulate the local economy as the economic recession turned into the Great Depression. The real estate and building interests lobbied for an Olympic Village. Initiated as a building contract, the concept became part of the invented tradition of the modern Olympics, and now all Summer Games consider an Olympic Village as part of the assumed infrastructural investment.

In the post-Second World War era, before the advent of substantial revenues from major corporate sponsoring and broadcasting rights, the cost of hosting the Games became a prohibitive factor in cities' bids. The number of cities bidding for the Summer Games began to decline. Only Los Angeles bid to hold the 1984 Games. The high cost of the 1976 Montreal Games exposed the stark contrasts between actual revenue and programmatic aspiration. On the one hand, corporate and broadcasting revenue was relatively small: only $35 million was received for the broadcasting rights. On the other hand, an expensive and lavish building campaign was undertaken. The net loss, including the cost of all infrastructural investments, has been estimated at $1,228 million (in 1995 US$; see Preuss, 2000: 244). Local and regional taxpayers made up the loss. Montreal marked a turning point of Olympian proportions.

The fiscal experiences of the Montreal Olympics made many cities think twice before putting in a bid. Again, Los Angeles charted a new course. As the only city short-listed for the 1984 Games, it had a strong negotiating position with the IOC. For the first time a special arrangement was estab- lished so that city taxpayers would not be responsible for any deficit. The US Olympic Committee set up a private non-profit corporation, the Los Angeles Organizing Olympic Committee (LAOOC), to make the arrangements so that the city taxpayers were not responsible for the costs. The LAOCC spent around $500 million renovating, rather than rebuilding, Olympic sites. They made $300 million from TV revenue and signed deals with thirty-four sponsors, paying between $4 and $15 million each, including Atlantic Richfield, which refurbished the Coliseum, AT&T, which set up telecommuni- cations, and General Motors, which provided cars. McDonald's paid for the swimming pool, and Coca-Cola paid $14 million to become the official drink of the 1984 Olympic Games. Total revenues amounted to $1,123 million, while costs were only $467 million. The LAOOC made a profit, corporations achieved global penetration as the Games were broadcast to 156 countries;

local businesses made money and the city became the center of world attention without the accrual of long-term costs or heavy debt burdens.

After the success of Los Angeles, the number of cities bidding for the Games increased. A new era of intense city competition to host the Olympic Games began. Table 8.3 lists the candidate cities for Summer Olympic Games since 1952. A number of trends can be noted. Candidate cities in the early period of this era were predominantly US, where cities were in much better shape after the Second World War and also had strong traditions of civic boosterism as well as political regimes dominated by the so-called urban growth machines. Detroit, for example, was a candidate four times between 1960 and 1972. Since the 1984 Games, however, and especially since 1988, when the fiscal experience of the 1984 Games was factored into Olympic bids, there has been an increase in the number of candidate cities and a wider range of cities outside of Europe and North America, which is part of the more general trend for many cities seeking to make the global connection (Hall and Hubbard, 1998; Roche, 1992; Short, 1999; Wilson, 1996).

Any city can make a bid to hold the Olympic Games. Bid cities are those that announce their intention to host the Games. Candidate cities are those that have been given approval by their National Olympic Committees, have submitted an application and have been short-listed by the IOC. For the 2008 Games ten cities made a bid: Bangkok, Beijing, Cairo, Havana, Istanbul, Kuala Lumpur, Osaka, Paris, Seville and Toronto. Five were short-listed in 2001 to candidate city status: Beijing, Paris, Toronto, Istanbul and

Table 8.3 Candidate and host cities, 1952–2008

Year	Host city	Candidate city
2008	Beijing	Toronto, Paris, Istanbul, Osaka
2004	Athens	Rome, Cape Town, Stockholm, Buenos Aires
2000	Sydney	Beijing, Manchester, Berlin, Istanbul
1996	Atlanta	Athens, Toronto, Melbourne, Manchester, Belgrade
1992	Barcelona	Paris, Belgrade, Brisbane, Birmingham (UK), Amsterdam
1988	Seoul	Nagoya
1984	Los Angeles	
1980	Moscow	Los Angeles
1976	Montreal	Moscow, Los Angeles
1972	Munich	Detroit, Madrid, Montreal
1968	Mexico City	Detroit, Lyon, Buenos Aires
1964	Tokyo	Detroit, Vienna, Brussels
1960	Rome	Lausanne, Brussels, Budapest, Detroit, Mexico City, Tokyo
1956	Melbourne	Buenos Aires, Mexico City, Chicago, Detroit, Los Angeles, Minneapolis, Philadelphia, San Francisco
1952	Helsinki	Los Angeles, Minneapolis, Amsterdam, Detroit, Chicago, Philadelphia

Osaka. Table 8.4 lists some of the basic characteristics of the short-listed bids as well as the main pros and cons identified by the IOC. Financial ability to host the Games and commercial "spin-off," i.e. the ability to use the Games to penetrate new markets, were the principal considerations. Given the willingness of the Chinese government to fund the Games and the huge commercial potential of the Chinese market, it was almost a certainty that Beijing would win. And win it did.

As globalization increases apace and as the benefits of the Games seem secure, bidding will only become more intense. Even within individual countries, competition between cities has increased. The US Olympic Committee (USOC), for example, had to consider serious bids from eight cities competing to become the US bid city for the 2012 Olympics. New York, Washington, San Francisco and Houston made the USOC'S first cut in October 2001. The four cities that failed to make the cut were Cincinnati, Dallas, Los Angeles and Tampa. There is renewed competition to host the Summer Olympics.

The hot competition to host the Olympics of 2004 was the backdrop to one of the few Olympic scandals to achieve widespread media coverage. The

Table 8.4 Candidate cities for the 2008 Olympics

City	Motto	Selling points	Main prospects	Main concerns
Beijing	New Beijing, New Games	Green, Hi Tech, Peoples Olympics	Offers huge new trade opportunities, government financial guarantees	Pollution and congestion, human rights issues
Istanbul	The Meeting of Continents	Crossroads of Asia and Europe	Location	Political and seismic instability, security concerns
Osaka	Sports Paradise Osaka	Green, community support	Japan is 2nd largest economy, venues already built	Concerns about Japan's economy
Paris	Oui	Financial guarantees from national government	Infrastructure in place, successfully hosted 1998 World Cup	Less commercial spin-off than Beijing
Toronto	Expect the World	Funding largely secured	Infrastructure in place	Less commercial spin-off than Beijing

structural context consisted of cities desperate to land the Games operating in a bidding system under which IOC delegates would visit bid cities, ostensibly to check out the site, and then vote in a secret ballot. Each IOC member held an all-important vote that they could cast in secret. Cities were lavish with their hospitality, and many delegates were eager to cash in on their voting power. There were no IOC checks on members visiting the cities or ethical guidelines in place. Bid authorities constructed dossiers of IOC members that highlighted their weaknesses. The dossier prepared for Stockholm's bid noted that Mohamed Mzali of Tunisia "should always be taken out for dinner on visits to Paris," while General Gadir of Sudan is "known for appreciating delicious food and drinking" (quoted in Calvert, 2002: 35). The system was endemically corrupt, with delegates leveraging more and more lavish gifts from cities' boosters who were increasingly eager to lubricate their bid with generous "hospitality." One IOC member was routinely referred to as the "human vacuum cleaner" for his ability to suck up money, gifts, holidays, flights and medical attention for himself and his family.

The scandal was revealed initially when a television station in Salt Lake City reported in November 1998 that the city's bid committee had paid for an IOC member's daughter to attend a private university in the USA. The next month a Swiss IOC member, Marc Hodler, stated in public that he believed that there was "massive corruption," and as many as twenty-five IOC members had their votes bought by bid cities. An IOC Commission was quickly formed in response to media pressure and especially to the worries of corporate sponsors that their products would be tainted with the label of corruption. The Commission reported in 1999 that seven IOC members were to be expelled, including Jean-Claude Ganga of the Congo, General Gadir of Sudan, Sergio Fantani of Chile and Augustin Arroyo of Ecuador. Ten more IOC members were warned about their behavior, two were exonerated. The scandal brought the scrutiny of media attention to the lavish lifestyle of IOC members who, it was revealed for the first time, were given business class flights to the Games and full-time drivers and luxury cars during their stays. They were installed in luxury accommodations with specific requirements for fresh flowers to be placed in their rooms each day of their residence. These requirements are in heavy contrast to the more spartan accommodations and basic conditions afforded to the athletes of the Games. The penchant for President Samaranch to be referred to as "His Excellency" was symptomatic of an unelected, unaccountable para-state grown heavy and fat with excessive entitlements. The IOC members expected and were granted the privileged lifestyle of the global elite.

In the wake of the Salt Lake scandal, the IOC created an Ethics Committee and made changes to the bid process. In 1999, a new procedure was set up in which the guiding principle was, in the words of the IOC, "no visits, no gifts." This directive was less concerned with democratizing the process or even making it more transparent; it was essentially a response to the bad publicity and the fear of corporate withdrawal. As the head of McDonald's

German subsidiary noted, "If the corruption suspicions are confirmed, McDonald's will ask itself if sponsorship of the games still has a place on the group's image" (Korporaal and Evans, 1999).

The scandal was one sign of the increasing competition for cities to host the Games. There were obvious rewards in hosting the Games. The LA Olympics had shown that money could be made and that city residents would not have to face a heavy tax burden. Quite the reverse, in fact: major infrastructural investments could be paid for by the Games. The Olympic Games was the agent of a huge multiplier effect, bringing money, jobs, facilities and recognition to the city. The Games was a way to become more globally connected through the tangible legacy of facilities and improved air, road and rapid transit connections and the less tangible – and hence deeply attractive – notion of greater international recognition. As the world shrinks, it becomes ever more important for cities to have a positive image in a world of global flows. The image can attract business, tourists and institutions not just for the Games but also for some time afterwards.

THE URBAN IMPACT OF THE GAMES

In one of the most exhaustive studies of the Summer Games, Holger Preuss (2000) argues that hosting the Games, at least since 1972 and especially since 1984, has meant improved infrastructure and increased income and employment. Estimates as to the precise economic impact of the Games are difficult to assess, but a report by McKay and Plumb (2001) estimates that the net economic impact on the host city averages from $4 billion to $5 billion. Barcelona was the exception with a net impact of $16.6 billion.

The biggest winners of the Olympic windfall are the political regimes running the city that have the opportunity to reshape the city's desired image (Andranovich *et al.*, 2001). The metanarrative is the creation of the global city, well connected to the outside world, presenting a positive image to millions of viewers around the world, all potential visitors, tourists and investors. The positive image presented is one of modernity and multiculturalism, part of the shared global discourses of democracy and liberalism while also adding a touch of the uniquely local. A distinctive place connected to a shared global space. A local and global package is presented that plays to global connections and local distinctiveness.

To the metanarrative of creating a global city, or at least the image of the global city, there are more subtle place-specific discourses. Thus for Seoul, it was the opening of the city and economy to the outside world; for Barcelona and Atlanta, regional economic development was important; while international positioning of the city was a key element in the Sydney Games.

Real estate and building companies also benefit from the extra business and contracts. There is a multiplier effect that filters down the income hierarchy. The Games in many ways acts as a Keynesian pump primer of the local urban economy.

Most of the negative costs are borne by the weaker groups in the city, especially those inhabiting prime inner city sites. The preparations for the Games have been associated with a spatial removal of the poor and the marginal, which can be both temporary and permanent.

Not everyone welcomes the Games. Lenskyj (2000) has compiled a list of the resistances to both hosting and bidding for the Games. Olympics have not gone unchallenged or uncriticized especially in societies where a democratic politics allows critics a platform for dissent.

The Games act as an important tool to literally reshape the city, in both discursive and spatial terms. Hosting the Games involves a massive restructuring of urban space. While an already rich minority benefit a great deal, and most benefit some, the poor and marginal tend to become poorer and more marginal. Hosting the Games creates the global city at the expense of the local weak. To this general picture must be added the difference between individual cities hosting the Games. In terms of house prices, for example, both Seoul and Barcelona saw a rapid rise in house prices while Atlanta and Sydney experienced very little increase. The course of urban renewal also varied. In Barcelona, for example, the restructuring turned the largely abandoned port area into a public space. The Games were used to modernize much of the city infrastructure. It was public-sector-led with most emphasis on public projects and the transformation of public spaces in the city (see De Moragas and Botella, 1995). In Atlanta, in contrast, it was private-sector-led with much little public infrastructure investment. The Atlanta Games were used by the local urban regime to push through major urban restructuring that primarily benefited business and corporate interests. While local people were employed during the Games, the reshaping of the city simply repeated and reinforced income and race cleavages. The destruction of 5,000 public housing units and the diversion of $350 million from redistributional spending on the poor to regressive spending for the Olympics displaced almost 15,000 people. Sydney, by way of yet another contrast, was a return to public sector dominance with major public sector investments in airport and road connections. The Games were held at a former abandoned waste site, Homebush Bay, and so the Games were part of a wider remediation project and greening of the city (Cashman and Hughes, 1998).

The Games also have a longer-term effect on the city. The city is permanently transformed by the spatial restructuring of hosting the Games. In Seoul the Chamsil area was redeveloped, in Barcelona the sea front was opened up, in Atlanta there was a central city gentrification that involved the construction of lofts, telecom hotels and high tech offices. In Sydney Home Bush Bay was cleaned up. In Athens it is planned to reclaim disused quarries, waste dumps and old army barracks. There is also the physical legacy of the Olympic villages constructed to house the athletes. In Seoul, Barcelona and Sydney, new Olympic villages were built and became new neighborhoods. The stadia constructed to host the Olympic events also became permanent sites able to be used for subsequent events. The Seoul

Olympic stadium was a venue for 2002 FIFA World Cup. In Atlanta the stadium became the permanent home of the Atlanta Braves baseball team. Perhaps the largest infrastructural legacy is the upgrade of airports, telecommunications, mass transit schemes and road networks.

Hosting the Games provides a significant opportunity to forge new and improved links with the wider world that plug the city more effectively into the global flows of capital, people and ideas. Amongst the positive effects of recent Games are the clean-up of derelict and abandoned sites, the creation of attractive public spaces, legacies of new spaces and improved athletic facilities and even an increase in the environmental quality of the city. The improvements in the air quality in Beijing, part of the city's Olympic bid package, will be enjoyed by all the residents of the city. As with most urban social changes there is an unequal distribution of costs and benefits. The poor tend to gain least and pay more. However, for every Atlanta there is also a Barcelona and Sydney where there were general improvements to the quality of urban life that were not entirely regressive. While all cities are rewritten by hosting the Olympics, the particular narrative can vary enormously.

Since their inauguration in 1896, the Summer Olympics have become increasingly global, with both athletic participation and television viewing truly international phenomena. The national level, often considered increasingly redundant in many discussions of globalization, is shown in this study to be alive and well. The Coubertin project was to promote international harmony through sports competition organized at the national level. Nations were not superseded by the Games, they were integral and essential. The Olympic Games celebrate national identity. At the city level the hosting of the Games provides one of the most obvious cases of glocalization as the global spectacle is centered in a specific city, while the hosting of the event tends to reconnect the city into a global space of flows.

The Summer Olympics are an example of cultural and economic globalization. The number of countries competing has steadily grown until it covers most of the globe. However, both athletic success and the hosting of the Games reflect the global inequalities in wealth. Because of the increasing corporatization of the Games, they also have become a launching site for economic globalization as major corporations use the Games as a platform for global penetration. The Games provide the opportunity for firms to sell their goods and services to a global market. The hosting of the Games in Beijing was in large measure driven by the need to find an entry into the vast Chinese market. The Games have become an important process of globalization, but a process that reinforces rather than undermines nationalism and national identity.

The Games have become an important transformative experience that denotes and connotes global cities. The globality of a city is never given or fixed, or even the result of some inert attribute; it is enacted, performed, spectacularized. This is evident in the bidding and hosting of the Summer Games. They have become the mega-event with the ability to create, reinforce and consolidate global city status.

9 The super-rich and the global city

There has always been inequality in the distribution of wealth and income at both a national and global level. But what is becoming increasingly evident is that the wealth of the world is becoming more unevenly distributed, with the rich becoming very rich and the poorest segments of the world's population possessing a diminishing share of the world's wealth (see Bergesen and Bata, 2002; UNDP, 2002). The top 0.25 percent of the world's population have as much wealth as the other 99.75 percent.

The tiny elite of very rich people have prospered in the globalized economy. In 2002 there were an estimated 7.2 million people with investable financial assets of more than one million dollars. We will refer to this group as the *super-rich*, specifically defined as high net worth individuals who have financial assets of more than $1 million. This group wields enormous financial and political power yet they are rarely subjects of study (exceptions being: Haseler, 2000; Lundberg, 1988; Smith, 2001), despite the fact that there is a pressing case for rigorous and methodical exploration of the "super-rich." The stock-in-trade of the policy-oriented social scientist is the "deserving" poor, whose living and housing conditions have been documented *ad infinitum*, and a more recent concern with the new middle class. Pahl (2001) suggests that some researchers feel an obligation to reach out to those less advantaged in society, they feel they can make a difference to their lives and (crucially) they find it easier to obtain research funding for work on poverty. In contrast, it appears difficult to gain funding for research on the super-rich, with research councils and foundations understandably uncomfortable about work that might raise questions about the way the elite amass wealth. There is also the issue of positionality in social science research. The middle class researcher finds it easier to gain access to the poor and middle class. The poor are more easily studied than the rich. Academic researchers are always in a position of relative advantage over the poor. This power asymmetry privileges the academic researcher. Access becomes more difficult the further up the income scale as the power relationship shifts from researcher to subject. The rich and wealthy, in contrast to the poor, are neither so deferential nor accessible. It is easier and more convenient, with the added lustre of "doing good," to study the poor rather than the rich:

> Whilst it is relatively easy to measure, analyse, and report on the form and pattern of the bruises in society, it is equally easy to ignore the fist that makes them. If we spend all our time on easily-funded bruise studies, we may be colluding with those concerned to keep the power of the fist unchanged.
>
> (Pahl, 2001: 883)

The lack of academic scrutiny of the super-rich makes it difficult, if not impossible, to make meaningful conclusions about the processes that allow some to live in luxury and others to wallow in poverty.

The disparity between the incomes of "high net worth individuals" and the majority of the world's population should not be surprising given the existence of a capitalist world system that thrives on inequality and exploitation. Nonetheless, commentators on both the Left and Right have remarked on this growing inequality, suggesting it raises serious questions about social cohesion (Castells, 1989; Hutton, 1995; Pahl, 2001). But neither the market nor the redistributive policies pursued by (some) nation-states seem able to prevent the rise and rise of the "super-rich."

The super-rich have the power and money to manipulate the state and market to their own ends, often positioning themselves beyond the jurisdiction of nation-states (e.g. exploiting tax loopholes in off-shore havens including the Cayman Islands, Channel Islands and Luxembourg). But even on-shore trading has become more attractive for the wealthy as personal and corporate tax rates have been cut in most countries of the world. The established financial centers of London, New York, Singapore and Frankfurt have become increasing adept at serving the needs of high net worth individuals with market deregulation offering enhanced investment performance.

THE SUPER-RICH AND THE GLOBAL ELITE

The super-rich are even invisible in most analyses of the emerging global elite. It is the professional and managerial class who are most often depicted as the beneficiaries of global restructuring. In describing the characteristics of the global elite that ensures the hegemony of the world capitalist system, Sklair (1997) suggests four main fractions:

- multinational corporate executives and their local affiliates;
- globalizing bureaucrats;
- globalizing politicians and professionals;
- consumerist elites (merchants and media).

Of course, there is much overlap between these categories, each of which plays a distinct role in the perpetuation of the global capitalist system. These groups are the knowledge-rich (and asset-rich) individuals charged with managing the global economy. Yet to suggest that these fractions are the principal beneficiaries of globalization is to suggest the key social divide in contemporary society is between a *service* class, conceptualized as those in key managerial positions (particularly in the advanced producer services), and a *servicing* class of low-paid, low-skilled workers. This is the divide that preoccupies Castells (1989) when he talks of the increased differentiation of labor within two equally dynamic sectors of the economy: the information-based formal economy (characterized by white, university-educated workers) and the "downgraded labor-based informal economy" (associated with mini-mally educated ethnic minorities). To demonstrate the existence of such occupational polarization, Castells assembles a wealth of data from New York and Los Angeles showing the rising gap between incomes in the knowledge-rich and knowledge-poor workforce. But a key divide that should concern us is not just that between the work-rich and work-poor (the divide that so preoccupies many writers on gentrification and urban restructuring) but between the super-rich and the rest.

It is important to distinguish the "super-rich" as an important separate category of the global elite. One of the earliest acknowledgements of the exis-tence of the super-rich was Thorstein Veblen's (originally 1899, reprinted 1985) *The Theory of the Leisure Class*. Writing at a time of increased inequality and the amassing of personal fortunes in the USA, Veblen described a group of people he described as the leisure class whose wealth freed them from the need for employment. Their consumption patterns were not so much con-nected to basic needs, as structured by the need to create what he described as a "decorous" appearance. The leisure class was thus recognizable not by occupation or even ownership of the means of production but by their patterns of consumption. Of particular importance was conspicuous con-sumption, whereby the super-rich signalled their exemption from the need to work. Examples of ostentatious consumption include the construction of large holiday mansions and villas; the acquisition of arcane and redundant artefacts; an interest in elite sports; and a tendency to accumulate a staff of servants, manservants, groundskeepers, grooms and the like to further signal their distance from the world of work. This marked them out as a distinctive group (as F. Scott Fitzgerald remarked, "They are different from you and me," Ernest Hemingway reportedly replying, "Yes – they have more money"). Veblen consequently regarded such acts of conspicuous consumption as essentially wasteful and excessive, a symbol of a pecuniary wealth distanced from the world of work. Veblen's acerbic analysis presages Bourdieu's work on the importance of cultural capital as a means of distinction (Bourdieu, 1984).

Veblen's analysis sets out some useful markers for looking at today's wealthy. While patterns of consumption may have changed, there is still the same need among the super-rich to create an appearance: while there may be very wealthy individuals who lead lives of studious asceticism, the majority continue to conspicuously consume. In this sense, their lifestyles are "on display" to the rest of the world. While academic analysis may shun them, the super-rich preoccupy the popular press who scrutinize their consumption down to the tiniest details. This does not mean that the less affluent are incapable of developing their own leisure and consumption patterns, or that all people aspire to the lifestyles and fashions of the super-rich. Nonetheless, the super-rich play an important symbolic role in the perpetuation of global capitalism, showing that it creates the possibility of a form of consumption based on profligacy and seemingly infinite choice.

We think it important to add a fifth group to Sklair's (1997) classification of the global elite: the global super-rich. After all, it is this group who has chiefly benefited from globalization to amass even more personal wealth. They are key movers and shakers in the global economy and can be conceptualized as a global ruling class whose reach and influence encompasses the entire capitalist system. The super-rich need to be considered as pivotal to the articulation of the global economy. This particular elite, in effect, sets the world economy in motion: an economy that is subsequently attended to by a "mass affluent" service class (including those who provide financial services for the super-rich). In turn, this managerial service class are serviced by a lower-paid, servicing class.

LOCATING THE SUPER-RICH

The neglect of the super-rich by the social sciences is in part caused by the lack of data available for "locating" them. It has been a long-standing criticism of official government statistics that there is inadequate information about income. More widely known and potentially more useful for locating the super-rich than government publications are the lists compiled by business monitors and newspapers, notably the *Sunday Times* Rich List (which lists the 1,000 richest individuals in UK) and the *Forbes 400* (drawn up by the US magazine targeting the elite business community). These lists are complemented by a multitude of lists documenting the hundred richest individuals in Europe, the fifty richest individuals in Latin America and so on. While the way these estimates of net worth are compiled leaves some margin for error, they are a useful barometer of the levels of wealth accruing to the very rich.

Table 9.1 represents a best estimate of the world's fifty richest individuals, as compiled by *Forbes* magazine. From this, one can begin to obtain a picture of the "geography" of the super-rich. Perhaps unsurprisingly, it is the high-income economies of the West which provide most names on the list, with North America accounting for 22 of the 50, and Europe for another 13. Beyond this global core, the majority are drawn from the oil-rich states of the Middle East; there are no names on the list from Africa, South America or Australasia. In many ways then, the geography of the super-rich seems to mirror the unequal share of wealth between the world's nations, with the larger high-income economies boasting the greater share of the world's super-rich.

Table 9.1 The world's richest individuals

			Position	2000 wealth	2001 wealth
1	Robson Walton*	USA	Retailing (Wal-Mart)	$45.3bn	$52.8bn
2	Bill Gates	USA	Software (Microsoft)	$37.5bn	$53.1bn
3	Larry Ellison	USA	Computers (Oracle)	$29bn	$8.1bn
4	King Fahd*	Saudi Arabia	Oil	$20bn	$17.5bn
5	Warren Buffett	USA	Investments	$17.3bn	$19.3bn
6	Paul Allen	USA	Software (Microsoft)	$17bn	$25bn
7	Sheikh of Abu Dhabi	AUAE	Oil/investments	$15.3bn	$12.5bn
8	Forrest Mars Jr*	USA	Confectionery	$14bn	$10.3bn
9=	Karl abd Theo Albrecht	Germany	Supermarkets	$13.3bn	$8.5bn
9=	Prince Alwaleed	Saudi Arabia	Investments	$13.3bn	$9.3bn
9=	Barbara Cox Anthony and Anne Cox Chambers	USA	Media	$13.3bn	$12.1bn
12	Emir of Kuwait	Kuwait	Oil/investments	$12bn	$10.6bn
13=	Sultan of Brunei	Brunei	Oil	$10.6bn	$18.7bn
13=	Kenneth Thomson	Canada	Media/oil	$10.6bn	$7.4bn
15	Steve Ballmer	USA	Software (Microsoft)	$10.5bn	$14.3bn
16	Liliane Bettencourt	France	Cosmetics	$10.1bn	$8.6bn
17	The Bass Family	USA	Oil	$9.7bn	$8bn
18	Pierre du Pont*	USA	Chemicals	$9.3bn	$8.1bn
19=	Philip Anschutz	USA	Oil/railroads	$8.6bn	$6.8bn
19=	John Kluge	USA	Media/phones	$8.6bn	$6.8bn
21=	Silvio Berlusconi	Italy	Media	$8.5bn	$5.1bn
21=	Johanna Quandt*	Germany	BMW cars	$8.5bn	$6.1bn
23	Bernard Arnault	France	Luxury goods	$8.4bn	$3.7bn
24	Sheikh Makhtoum	UAE	Oil/finance	$8bn	$7.5bn
25	Michael Dell	USA	Computers	$7.8bn	$12.5bn
26=	Lukas Hoffman*	Switzerland	Pharmaceuticals	$7.6bn	$10.6bn
26=	Leo Kirch	Germany	Media	$7.6bn	$3.1bn
28	Li Ka-shing	Hong Kong	Property	$7.5bn	$7.9bn
29	Abigail and Edward Johnson	USA	Investments	$7.4bn	$6.9bn
30=	Gérard Mulliez*	France	Retailing	$7.3bn	$6.1bn
30=	Robert and Thomas Pritzker	USA	Investments	$7.3bn	$6.8bn
32	Ted and Norman Waitt	USA	Computers	$7.1bn	$4.8bn
33	Gordon Moore	USA	Microchips	$7bn	$9.3bn
34=	The Mellon Family	USA	Banking	$6.6bn	$6.2bn
34=	Samuel and Donald Newhouse	USA	Publishing	$6.6bn	$5.6bn
34=	Sumner Redstone	USA	Media	$6.6bn	$5.8bn
37	The Haniel Family	Germany	Pharmaceuticals	$6.3bn	$7.7bn
38	Yasuo Takei	Japan	Finance	$6.2bn	$4.8bn
39	Ernesto Bertarelli	Switzerland	Pharmaceuticals	$6.1bn	$2.4bn
40	The Kwok Brothers	Hong Kong	Property	$6bn	$6bn
41	Donald Fisher*	USA	Retailing	$5.9bn	$5.1bn

Table 9.1 (continued)

		Position	2000 wealth	2001 wealth
42= Lee Shau Kee	Hong Kong	Property	$5.7bn	$6.9bn
42= The Seydoux and Schlumberger Families	France	Media/oil/textiles	$5.7bn	$4bn
44 The Rockefeller Family	USA	Oil	$5.6bn	$5bn
45= Walter Haefner	Switzerland	Car sales/software	$5.5bn	$4.5bn
45= Ted Turner	USA	Media	$5.5bn	$4.3bn
47= Suliman Olayan*	Saudi Arabia	Investments	$5.3bn	$4.4bn
47= Stephan and Thomas Schmidheiny	Switzerland	Cement	$5.3bn	$4.8bn
49 François Pinault	France	Retailing	$5.2bn	$4bn
50 Stefan Persson	Sweden	Retailing	$5.1bn	$4.3bn

Source: *Forbes 400*, October 2000; *Forbes Global*, July 2000 .
Note:
* Indicates estimate of family wealth.

To provide a convincing analysis, it is obviously insufficient that we study the world's richest thousand, let alone the richest fifty; the global capitalist class is now much larger. Moreover, the idea that the super-rich dwell within a specific nation-state is a clearly outdated and flawed assertion. Instead, the global elite (i.e. both global managers and the global super-rich) must be regarded as transnational. Three main reasons: first, its members tend to have an outward-oriented global outlook rather than inward-oriented national perspectives on a variety of issues. Nowhere is this truer than in the global outlook of TNC executives and CEOs, where key players inevitably have an interest in monitoring and intervening in the international business environment. This is well illustrated in the involvement of members of the global capitalist class in the efforts to secure the Free Trade Agreements that serve to benefit their particular business interests. Each fraction of the global capitalist class thus sees its mission as organizing the conditions under which its interests and the interests of the system as a whole (which usually coincide) can be furthered in the global sphere rather than the local.

Second, it is evident that while members of the global elite may be over-represented in the core nations, they often consider themselves "citizens of the world" as well as of their places of birth. Leading exemplars of this phenomenon listed by Sklair (1997) include Jacques Maisonrouge, French-born, who became in the 1960s the chief executive of IBM World Trade; the Swede Percy Barnevik who created Asea Brown Boverei, often portrayed as spending most of his life in his corporate jet; the German Helmut Maucher, CEO of Nestlé's far-flung global empire; and the "legendary" Akio Morita, the founder of Sony. These examples serve to make the crucial point that state-centered statistics are wholly inadequate for documenting the geographies of the global elite: residence in a particular nation-state or possession of national citizenship does not mean that the individual concerned has any particular allegiance to that nation, or that the individual has been brought up within a particular business culture.

Third, it is apparent that the global elite must be regarded as transnational to the extent that they share similar global lifestyles. For Short and Kim (1999) the lifestyles of global managers present perhaps the clearest evidence that the shared consumption of similar goods and images is resulting in the creation of global lifestyles. The mobility of global managers is, however, eclipsed by that of the global super-rich. They are able to move easily from nation to nation by executive jet (rather than travelling by business class); they stay only in five-star hotels; they are able to access exclusive clubs and restaurants, they frequent ultra-expensive resorts in all continents, and collect the objets d'art which can only be obtained in the most exclusive shops and auction houses. In short, their space-time routines center on a globally diffuse set of spaces regarded as "the right places to see and be seen," a fast world that could not be more different to the experience of those resigned to living in a world that appears increasing slow in comparison (Urry, 2001).

The global super-rich are a vital if little studied part of a new global elite. This demands that we take their transnational lifestyles seriously, and think about how their "global reach" is implicated in the making of new global orders.

Considerable attention is devoted to the heightened salience of migration in the contemporary world, with a particular focus on "transnational communities." Hannerz (1993) describes four groups of people who fall within this definition, forming communities which may be described variously as diasporic, hybrid or nomadic:

- transnational business people, including the high-waged, highly skilled professional, managerial and entrepreneurial elites usually associated with finance, banking and business services (see Beaverstock and Boardwell, 2000);
- Third World populations, comprising low-waged immigrants who occupy insecure niches in the unskilled or semi-skilled sectors of the urban service economy;
- expressive specialists, who participate in the cultural scene in areas such as art, fashion, design, photography, film-making, writing, music and cuisine (Scott, 2000);
- tourists, whose transnational status is often ephemeral but who make up a major proportion of those who are living outside their "home" space (Urry, 2001).

We might then add another transnational community to Hannerz's list – the super-rich – who must be regarded as quite distinct from any of the four categories identities described by Hannerz, yet share aspects of their transnational lifestyle.

The existence of such transnational communities highlights major deficiencies in those theorizations of social change that assume they occur on a state-by-state basis. This "embedded statism" (Taylor, 2000) is a key obstacle to a full theorization of the super-rich as a transnational community. Here, Manuel Castells' explanatory theory of the "network society" offers a more convincing analytical framework for exploring the geographies of the super-rich. Based on his tentative identification of the network as the key social form organizing people in relationships of production, consumption, experience and power, Castells' idea is that we live in a network society, reproduced unevenly by "flows of capital, flows of information, flows of technology, flows of organizational interaction, flows of images, sounds and symbols" (Castells, 2000: 418). The global city network provides the hubs and nodes around which the first layer of flows is organized. The network society has three layers: the electronic impulses in networks, the places which constitute the nodes and hubs of networks, and the spatial organization of cosmopolitan elites who perform key translation tasks in these networks. Yet both Hannerz and Castells downplay the importance of the super-rich in commanding and controlling the world system, instead suggesting the key actants in world city networks are the service class. While the latter are certainly of significance in shaping the global space of flows, one might legitimately question their global outlook and transnationalism. In Castells' view, these service elites are *"global,"* unlike ordinary people who are deemed to be "local." But Beaverstock (2002) suggests the designation of service elites as truly global may be misleading, with many transnational elite workers forming "personal micro-networks" that center on the expatriate residential and leisure-orientated spaces characteristic of major world cities. In contradistinction, the super-rich inhabit a truly global set of spaces, often owning multiple residences, holiday homes and retreats. As such, they are crucial in the articulation of the space of flows, their truly global presence and outlook fulfilling a *de facto* governance function (not least through the direct and indirect influence they might have on the charities, foundations, business organizations, think tanks, advisory councils and corporate committees who influence fiscal and social policy at a global scale, not to mention the World Bank, IMF, OECD, WTO and agencies of the UN). Perhaps more so than any other group, it is the super-rich who shape the networks that lie at the heart of the space of flows.

The global super-rich need to be conceptualized as a truly transnational and cosmopolitan fraction of the global elite whose activity spaces are pivotal to the articulation of global flow. Bauman (2000) suggests that the fundamental consumption cleavage in contemporary society is between these "fast subjects" who dwell in transnational space and those "slow subjects" whose lives remain localized and parochial. An important divide in contemporary society is that between cosmopolitans (who dwell in the space of flows) and the rest (who dwell in a world of places). There is both a two-speed city and a two-speed world. The fast world is one consisting of airports, top-level business districts, top of the line hotels and restaurants, chic boutiques, art galleries and exclusive gyms – in brief, a sort of glamor zone that is fundamentally disconnected from the life worlds of the vast majority of the world's population (see also Sassen, 2000). *Forbes* magazine lists (presumably without irony) the commodities that the super-rich "need" (Table 9.2). The global super-rich remain as much defined by their consumption patterns and cultural habitus as their wealth per se, with the acquisition of a number of icons of global savoir-faire (e.g. British motor cars, French wine, Italian clothing) marking off the super-rich as truly transnational (Short and Kim, 1999). An advertisement in the *New Yorker* magazine in September 2002 for the St Regis Hotel had as advertising copy, *At home with Swiss banking, French couture, Italian racing and the St Regis New York.*

The commodities consumed by the super-rich, though varied in origin, are characteristically purchased/consumed in global cities. It is particularly significant that these key sites of consumption are predominantly located in London and New York, pointing to the particular significance of the "NY–LON" nexus in the lives of the global super-rich. Nonetheless, it is too simple to suggest that the consumer geographies of the super-rich simply map onto the cities that are the acknowledged centers of the global financial system, with high-status tourist- and consumer-oriented spaces like Mustique, Monaco, St Moritz and Cannes obviously playing a crucial role as playgrounds of the rich and famous (despite their evidential lack of importance as centers for the advanced producer services). In such cases, the primary lure of these spaces for the global rich is clearly conspicuous consumption, but this does not mean that these are spaces are insignificant in the articulation of global flows of all kinds.

While much of transnational consumption may occur in major global cities, it is clear that the distinctive consumer habits of the cosmopolitan super-rich are played out in spaces that share certain characteristics. Central here is the air of exclusivity that is associated with the spaces of the super-rich. Given the fact that super-rich consumption is preoccupied with the importance of sign rather than use values, it is the space as much as the commodity that is consumed by the super-rich. An emphasis on style, pre-sentation, and performance is part and parcel of the leisured consumption of the super-rich. They may live in the "fast world," but the cultural tastes of the new cosmopolitans do not extend to fast food. When they eat at a restaurant, it is not so much a meal that is being consumed, but the experi-ence of eating out in a space that can be best described in dramaturgical terms (i.e. where staff follow a carefully determined script and play out their choreographed roles in a "theater of dining"). Bell and Valentine (1996) argue, in consumer society "you are where you eat." For example, it is illustrative that the super-rich eat in La Tour d'Argent, not McDonald's; that they shop on Rodeo Drive, not the discount stores of South Central LA. The super-rich consumption occurs in spaces that, by and large, exclude those who disturb the ambience of affluent, leisured consumption. This means that

Table 9.2 Items needed by the super-rich

Product	1976 price	2000 price	2001 price
Coat/natural Russian sable, Maximilian at Bloomingdale's	40,000	195,000	195,000
Silk dress/Bill Blass Ltd., classic	950	3,000	3,090
Loafers/Gucci	89	335	350
Shirts/one dozen cotton, bespoke, Turnbull & Asser, NYC	448	2,700	2,820
Shoes/men's black calf wing tip, custom-made, John Lobb, London	202	2,328	2,544*
School/preparatory, Groton, one-year tuition, room, board	4,200	28,620	30,340
University/Harvard, one-year tuition, room, board, insurance	5,900	33,110	34,269
Catered dinner/for 40, Ridgewell's, Bethesda, MD	2,200	5,036	5,400
Opera/two season tickets, Metropolitan Opera, Saturday night, box	480	5,000	5,300
Caviar/beluga malossol, 1 kilo, Petrossian, Los Angeles, CA	283	2,700	2,700
Champagne/Dom Perignon, case, Sherry-Lehmann, NYC	300	1,500	1,500
Filet mignon/7 pounds, Lobel's Prime Meats, NYC	50	189	189
Dinner at La Tour d'Argent/Paris, estimated per person (including wine and tip)	34	203	208*
Piano/Steinway & Sons, concert grand, Model D, ebonized	13,500	83,100	86,100
Flowers in season/weekly arrangements for six rooms, Christatos & Koster, NYC, per month	1,400	7,215	7,215
Sheets/set of lace linen, Pratesi, queen-size	1,218	3,360	3,460
Silverware/Kirk Steiff Co., Williamsburg, shell pattern, four-piece place setting for 12	1,341	4,680	4,680
Hotel/two-bedroom suite, park view, The Stanhope Park Hyatt Hotel, NYC	333	1,090	1,090
Face-lift/USA Academy of Facial Plastic & Reconstructive Surgery, NYC	4,000	11,000	12,500
Hospital/VIP, Washington, DC. Hospital Center, one day, concierge, security, gourmet meals	325	1,424	1,481
Psychiatrist/Upper East Side, NYC, 45 minutes, standard fee	40	220	240
Lawyer/established NYC firm, partner, estate planning, average hourly fee	80	555	595

Table 9.2 (continued)

Product	1976 price	2000 price	2001 price
Spa/The Golden Door, CA, basic weekly unit	1,250	5,375	5,725
Perfume/1 oz. Joy, by Jean Patou	100	380	400
Sauna/US Sauna & Steam Co., 8×10×7 feet, eight-person, cedar	5,000	12,200	12,810
Motor yacht/Hatteras 75	214,700	3,009,200	3,107,000
Sailing yacht/Nautor's Swan 68	384,300	2,000,000	2,000,000
Shotguns/pair of James Purdey & Sons, Griffin & Howe, NYC	20,000	105,487	114,686*
Thoroughbred/yearling, average price, Keeneland summer select sale	67,300	621,015	710,247
Swimming pool/Olympic (50 meters) Mission Pools, Escondido, CA	180,000	900,000	963,000
Tennis court/clay Putnam Contracting, Inc., Plainville, CT	25,000	55,000	55,000
Train set/Blue Christmas G gauge, LGB, at FAO Schwartz, NYC	178	550	499
Airplane/Learjet 31A, standard equipment, certified, ten passengers	1,800,000	6,419,600	6,525,600
Helicopter/Sikorsky S-76C+, full executive options	1,300,000	8,250,000	8,450,000
Automobile/Rolls-Royce Silver Seraph	38,000	219,900	229,990
Airline ticket/British Airways Concorde, round-trip NYC–London	1,512	11,926	12,284
Telephone call/ten minutes, AT&T, NYC–London	12	14	14
Cigars/Aniversario No. 1, Dominican Republic, 25 cigars, Davidoff, NYC	186	613	650
Magazine/*Forbes*, one-year subscription	15	60	60
Duffel bag/Louis Vuitton, Keepall Bandoulière, 55 centimeters	739	710	700
Watch/Patek Philippe classic men's gold, leather strap	2,450	10,800	10,800
Purse/Hermes, 'Kelly Bag,' calfskin, rigid, 28 centimeters	550	4,700	4,700
	Total 1976	*Total 2000*	*Total 2001*
	4,118,665	22,019,895	22,605,236

Source: http://www.forbes.com/2001/09/27/400.html

Notes:

* Based on currency exchange rates as of August 31, 2001.

All prices in US dollars.

the super-rich may rest, work and play in global cities, but that they rarely have contact with the ordinary dwellers of the city: the metaphor of "street people" versus "air people" appears particularly appropriate here, with the super-rich both physically and psychically distanced from the everyday rituals that are played out on the city's streets (as well as the dangers of urban life which are inevitably associated with street life).

While the super-rich may move through through global cities, their cosmopolitan practices and lifestyles rarely break out of the exclusive transnational spaces which stand at the intersecting points of particular corporate, capital, technological, information and cultural lines of flow. These transnational spaces have profoundly marked the urban landscape of global cities, transforming their social morphology in the process, so that districts like Knightsbridge, the Upper East Side or Passy d'Auteuil may effectively be described as super-rich enclaves (even if the super-rich don't dwell there for very long before moving on to one of their other homes). The super-rich have therefore made an immense claim on global cities and have reconstituted many spaces of global cities in their own image. Their claim to the city is rarely contested, even though the costs and benefits to cities have barely been examined (Sassen, 2000). In part, this is because these spaces are often woven into the fabric of world cities in ways that belie their exclusive nature. The rich do not exclusively dwell in carceral spaces (as a cursory reading of the work of the LA school might suggest). After all, the rich do not enjoy consuming in spaces that look like fortresses, where security is omnipresent and where the less affluent are forcibly removed by heavy-handed security guards. The exclusive nature of spaces of cosmopolitan consumption is generally more subtle than this, a logic of "look but don't touch," bequeathing a logic of self-exclusion.

It is the fact that super-rich spaces are often so visible and physically accessible to the palpably not-rich that makes them so intriguing as spaces of consumption. In many ways, the notion that cosmopolitan consumption thrives on this visibility reinforces the idea that super-rich lifestyles are conspicuous. Sometimes, this takes interesting forms; for example, when the super-rich make a play of their charitable works, donations to celebrity auctions and (remarkably) their claims to be green consumers (conspicuous asceticism?). Generally, however, there has been a spectacular rise in luxury consumption, with the consumption patterns of the global elite acting as a marker for those further down the income scale. Robert Frank (2000) describes the process as "luxury fever," as consumption expectations are ratcheted up all across the income scale. The global elite are pushing up people's expectations and assumptions. In the USA, for example, the average size of house has doubled, in square feet terms, in the past thirty years. In part it is a function of the positional nature of consumption. We consume in order to position ourselves relative to other people. Not only do the global elite raise the upper limit, everyone is thus forced to spend more just to keep up, but they also become the perceived benchmark. Juliet Schor's work, for example, shows that people are no longer keeping up with the people next door, but the people they see on television and magazines (Schor, 1992, 1998). In order to keep up with these raised consumption standards people are working harder and longer as well as taking out more debt. It is not so much keeping up with the Jones but "keeping up with the Gates." The increase in luxury consumption has raised consumption expectations further down the income scale, which in order to be funded has involved increased workloads and increased indebtedness. A focus on luxury consumption is an important element in understanding popular consumption.

One of the key roles of new cosmopolitans, it seems, is to place themselves in the global shop window as examples of the success of the capitalist world system – as both key actors and beneficiaries of the network society. It is the global super-rich who probably do most to secure the cultural hegemony of capitalism, standing as a visible sign of the desirability of global mass consumption. The multiple-mediated consumption spaces of the super-rich matter, both culturally and economically, and they stand as a projection of the power of the super-rich across the world.

TOWARDS A RESEARCH PROPOSAL

This chapter has suggested that has been a surprising lack of study of the world's global elite, especially the super-rich. This omission is particularly surprising given the widening disparity between the incomes of the super-rich and those of the rest. However, to date most attention has been devoted to a particular fraction of the global elite, who, while affluent, lack the global outlook and mobility of the super-rich. To elide global managers with the global super-rich is clearly dangerous. While the former are key actors in a global economy, charged with attending to the global flows that are channelled through the hubs and nodes of the global economy, we have to argue that the super-rich eclipse their importance. Contra Castells, we argue that it is this group, not global managers, that is truly transnational.

Given the rising disparity of income witnessed at a global level, there is now a pressing need for studies that explore how the super-rich are able to manipulate global networks to their own advantage, and to study how their 'micro-networks' serve to articulate the global space of flows.

There are a number of important research questions including whether existing definitions of the super-rich (e.g. financial assets of more than £1 million) are adequate, and whether the distinction between the global super-rich and global managers is sustainable empirically as well as theoretically. Ultimately, based on the premise that the super-rich are defined by what they consume, it may well be that answers to these questions can only be obtained by ascertaining where key spaces of cosmopolitan consumption are and examining how they are used by the super-rich. Investigating such issues raises a number of intriguing methodological challenges, with the mobile lifestyles of the super-rich obviously foreclosing a number of established research techniques (ethnography appears out of the question, for example). Similarly, it seems there is no reliable inventory of the world's super-rich (and, in any case, it needs to be conceded that a roll call of high-worth individuals could generate a list some six million names long if one accepts the definition of the super-rich as those with assets of £1 million plus). Given the difficulty of a subject-centered approach, a tentative and exploratory framework for research could focus on the consumption of particular goods and the consumption of particular spaces.

We do not imply an attempt to distill fluid trends and meanings into a set of rigid interpretations or permanent answers. Our emphasis rather is on circuits and flows, on chains of connection and ambivalent meanings, and in theorizing these connections not in linear terms but in terms of circuits of exchange with permeable boundaries that are subject to all manner of creative leakages. This approach might proceed in two stages:

1   The construction of a matrix of the 'world's best' commodities/spaces of consumption (for example, the world's fifty best hotels, restaurants, fashion stores, etc.) cross-tabulated with the spaces in which they can be consumed/purchased/visited. The sources of such information would be the media and press consumed by the super-rich, including business magazines (*Forbes*, *Business Age*), lifestyle magazines (*Wallpaper*, *Homes and Gardens*), restaurant guides (*Michelin*) and so on. This matrix could additionally entail a study of the global events (fashion shows, sporting events, art openings) that might be expected to be incorporated in the micro-networks of the super-rich, as well as data relating to the air travel of the super-rich (e.g. airports with executive facilities, number of private flights, etc.). The resulting matrix could be subject to multidimensional analysis to produce a mapping of the networks of the super-rich akin to those produced by GAWC to represent the world city network (Beaverstock *et al.*, 2000a, 2000b).

2   The second phase would take case studies of the consumption spaces that are tied into the micro-networks of the super-rich. In each case, the objective would be to explore the role of these spaces as spaces of display and exclusivity serving the transnational super-rich rather than an essentially localized population. This would include examination of the social rituals associated with those sites and a broader historical consideration of these spaces as sites of conspicuous consumption. This will involve interviews with the proprietors and staff of such venues, together with analysis of company histories and documentation (where available).

This focus on spaces of cosmopolitan consumption will allow the development of a unique perspective on the changing global significance of the super-rich, and, by demonstrating the overlap between these spaces and the global city network, will generate new insights into the cultural and economic significance of the super-rich in the network of global cities and the global space of flows.

10 The global, the city and the body

The debates about globalization are curiously disembodied. While the recent discourses of the body tend to ignore the cultural weight and economic impact of globalization, the debates on globalization fail to mention the importance of the body. One is so concentrated on the micro that it ignores the macro while the other, in focusing on the global, loses sight of the embodied local. In this chapter I want to focus attention on the discursive space that lies between the global, the urban and the body and to outline possible connections that may be worthy of exploration and further discussion.

Let me begin with an historical observation on the doctrine of macrocosm and microcosm, which posits a direct relationship between the body and the universe. This world-view long held sway in the western tradition until the modern era. Figure 10.1 is abstract design from a work of Saint Isidore of Seville (560–636) which was first printed in the fifteenth century and informed much of the commonsense view of the world until the eighteenth century. At the center of this diagram are *Mundus, Annus* and *Homo,* the Latin words for *World, Year* and *Man.* Concentric circles surround them. On the outer circles are the names for the four elements, seasons and four humors. A trellis-like design, reminiscent of a Celtic interlace, binds everything together seamlessly. This diagram summarizes the notion that the human body both incorporates and reflects the wider cosmos. The zodiacal figure that showed the relationship between the star signs and parts of the body was also an important part of medieval cosmology (Figure 10.2).

In this tradition, the cosmos is understood as an analogy of the human body, while the human body is understood as a copy of the cosmos. The body is both an encapsulation of the world and a key to understanding the world. There was an enduring, specifically embodied understanding of the world and a cosmic grounding for our selves.

For most Renaissance scholars and artists this strong connection between the body and the cosmos manifested as an intense interest in the workings, dimensions and representation of the human body. Leonardo da Vinci sketched hundreds of diagrams of the body's anatomy and physiology. One of the most recognized and reproduced icons of Renaissance art and science is his Vitruvian Man, a representation of the human figure as ideal form,

Figure 10.1 Macrocosm and microcosm (from *De responsione mundi et astrorum ordinatione*, 1471).

Source: Courtesy of the Library of Congress, Washington, DC.

marked by the squared circle (Figure 10.3). Leonardo covered an astonishing range of subjects, from astronomy to architecture and armaments. He left over 7,000 pages of notes and diagrams which reveal an inquisitive mind, a great talent and a constant observation, and a keen representation of the world around him. He was particularly concerned with form and structure, sketching hundreds of anatomical drawings of humans and animals as well as many living portraits. The human body was a major focus of concern and an important a site of representation. He wrote,

> So that we might say that the earth has a spirit of growth, that its flesh is the soil, its bones the arrangement and connection of the rocks of which the mountains are composed, its cartilage the tufa, and its blood the springs of water. The pool of blood which lies around the heart is the ocean and its breathing and the increase and decrease of the blood in the pulses, is represented in the earth by the flow and ebb of the sea; and the heat of the spirit of the world is the fire which pervades the earth

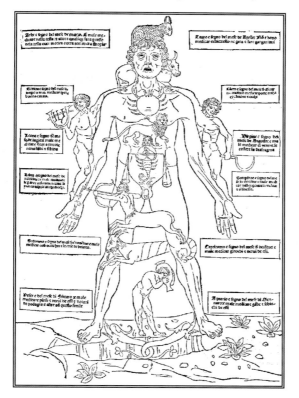

Figure 10.2 Zodiacal figure (from Ketham's *Fasciculus Medicinae*, 1493).
Source: Courtesy of the Library of Congress, Washington, DC.

and the seat of the vegetative soul is in the fires, which in many parts of the earth find vent in baths and mines of sulphur and in volcanoes.

(Richter, 1952: 45–6)

Leonardo expressed a common assumption of the time – the importance of the heart as the vital source of ebb and flow both in the microcosm and in the macrocosm. The link between macro and micro, the human body and cartographic representation was even more evident in other parts of Leonardo's voluminous notes. At one point he describes the order he will use in describing the human body:

Therefore there shall be revealed to you here in fifteen entire figures the cosmography of the "minor mondo" (the microcosms or lesser world) in the same order as used by Ptolemy before me in his Cosmography. And therefore I shall divide the members as he divided the whole, into fifteen provinces, and then I shall define the functions of the parts in every

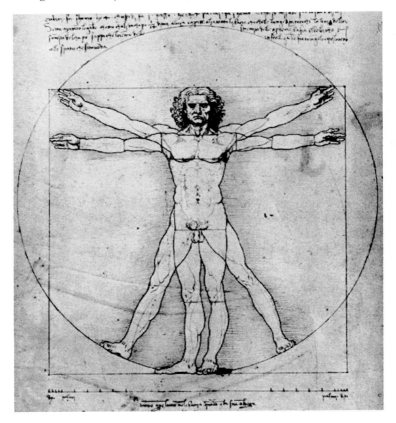

Figure 10.3 Vitruvian Man (Leonardo da Vinci, *Notebooks*, 1513).
Source: Courtesy of the Library of Congress, Washington, DC.

direction, placing before your eyes the perception of the whole figure
and capacity of man.

(MacCurdy, 1956: 161)

When we look again at Leonardo's famous figure of Vitruvian Man we now
see it in a new light: it is not only a new way of looking at the human body
but also a venerable view of representing the connection between the microcosm and the macrocosm.

In the Enlightenment project, the Cartesian dictum *Cogito, ergo sum* (I
think, therefore I am) lauded the thinking mind over the living body. Apollo
rather than Dionysus was idealized. Religion praised the world of the spirit,
and the Enlightenment praised the world of the mind, yet both of them
ignored or marginalized the living body; the quest was upward toward the
intellectual and spiritual realm and away from the grounded, earthly body.

And even in the psychoanalytic revolution of the twentieth century the body was a source of forbidden desires that a conscious ego and super ego tried to ignore and rationalize.

Because we have so recently inhabited such a disembodied world, it should come as no surprise that in recent years there has been revived interest in the body as a subject for analysis. The disembodied nature of many of our academic discourses, as well as the sensual revolution that made the body a more integral part of the presentation of self, has led to renewed concern with the human body. Following on from Foucault, there has been a tremendous emphasis on the body as an effect of arrangements of power and as a symbolic system which produces metaphors for power. But even here the Foucaultian body tends to a rather docile object, a site where power is acted on and out, rarely a place of active resistance.

How is the body the new measure of all things? What are some of the connections between this concern with the body as an object of analysis and our understanding of the global and the city? An obvious point, often missed in aggregate economic discourses, is that globalization is not just transference of cultures and beliefs, it is a movement of human bodies. Global flows quite literally involve the juxtaposition of different bodies from around the world in the shared spaces of global and globalizing cities.

Bodies are homes to viruses and disease. Global travel and movement allows the rapid spread of viruses. On November 1, 2002 a new disease, subsequently known as Severe Acute Respiratory Syndrome (SARS), broke out in China. By March 2003 the disease had spread to over thirty countries. By the end of April 2002, while the pneumonia-like illness had killed only 353 out of 5,462 reported cases, it had achieved global recognition and a global response. The World Health Organization (WHO) only lifted a week-long travel advisory warning regarding Toronto after the number of reported SARS cases fell and the Canadian authorities promised to introduce more effective screening at airports to limit the spread of the disease.

Quick, cheap and regular air travel links most parts of the world. The rapid and large flow of people around the world is creating global vectors of disease transmission. What used to take months or even years to move across human populations now can take weeks or even days. The establishment of global disease monitoring and reporting has also created a global health surveillance system and reinforced international organizations such as WHO. Global cities, with their many and varied global links, are transmission points for the international spread of contagious viruses.

Globalization involves not only the more acute awareness of body differences but also literally brings home the categories of bodily difference. The early European colonial adventures invoked sharp distinctions between native and non-native. The bodily forms of these differences became part of the discourse of colonialism and later of empire. In the nineteenth century these differences were reinforced by racist doctrines "proven" by the new sciences of anthropology and history. Differences in skin color and body shape became the embodied sources of division and difference overlain with value judgments and systems of power and repression. The other was so obviously the other when the body of the other was so different. Black skin, almond eyes, all became sources of difference. The native body could be whipped and beaten, the colonial body could be tethered and tied and the imperial body became the subjugated body. But there were also other narratives of difference. Colonial and imperial bodies also became loci of emotions and desires denied the urbanizing, industrializing, organizing bodies of the imperial center. The black man became a site of hypermasculinity, the female Asian a setting for a hyperfemininity. Black rage and Asian submissiveness became important colonial discourses that have an important legacy in the postcoloniality of many global cities. In many major cities in the USA, for example, massage parlors that sexually service male customers frequently tout the Asia connection. The image of submissive and sexually pliant Asian women is a key feature of the promotion, success and growth of massage parlors. The contemporary immigration of women from China, Korea and other Asian countries provide a steady flow of sex workers to fuel these postcolonial bodily imaginings.

The sexual body is now an important part of global flows. The flow of sex workers is an important element of labor migration, while sex tourism is a major form of international tourism. The international migration of sex workers has increased in recent years. The two driving forces are, first, economic incentives, as people try to make a better life for themselves and particularly their families, and, second, brute force. Many women are forced into the sex trade by blackmail and the threat of violence. The flow of sex workers across borders is increasing, as Thai women become prostitutes in Tokyo and Chinese women work in massage parlors in Los Angeles.

While the older colonial discourses still dominate and have a huge influence, the dominant narrative of the global and globalizing city is of the celebration of body differences. Black and white, Asian and non-Asian in some form of close connection have become the dominant forms of bodily portrayal. Global cities now contain a variety of body types; a rich variety of skin colors, body shapes, sizes and weights now form part of the visual texture of a truly global city. Cosmopolitan cities have a variegated stock of bodies. Marketing campaigns that use groups of people now try to display racial diversity. The United Colors of Benetton campaign, with its people of varied skin hue, is just one example of a global advertising imagery that echoes the embodied nature of global cities.

A personal note. You only notice something by its absence. I visited South Korea in 1993. I left the USA to visit a country that had a remarkable degree of cultural homogeneity. When I traveled on the metro system, even in the large city of Seoul, young children, less sanctioned than their parents for looking directly at strangers, would look at me as an obvious oddity. I was clearly not like them. I was a different body. Seoul's lack of cosmopolitan sensibilities was reflected in the gaze of the curious children.

Bodily differences are part of the global city and managers of globalizing cities realize that some degree of cosmopolitanism is an essential ingredient of becoming a truly global place. But there are tensions. Different bodies displayed on the screen or deployed in fictional narratives do not have the embodied sense of difference that bone and flesh do. And here we come across an interesting paradox. The least globalized cities often have the greatest fears of the foreign bodies. Part of the globalizing project is the acceptance and celebration of bodily differences. Bodies have become more important as a source of visual differences as clothes take on a more international look.

Our bodies are sites of globalization. We ingest food from around the world, wear clothes made and designed around the world. Hairstyles draw on global as well as local and national styles. In some cases our bodies are inscribed by globalization. Tattoos, for example, have become a form of global tribalist markings drawing on older, different traditions. Our bodies are increasingly described in more global terms, black and white, Asian and Non-Asian. Compare this to the names used by many premodern tribes: their names for themselves were often a translation of "we the people." Now we are described in more global, relativist terms. All of these differences are embodied in global cities where the global, the national and the local are juxtaposed. And the flows move up the scales as well as down. Local trends can become more globalized. Islamic fundamentalism, growing out of the dislocation of the Middle East, has become a global phenomenon with mosques in cities around the world now focal points for a global movement of religious expression. The dichotomy between cosmopolitanism and fundamentalism is often described by the clothes we wear, the lack or presence of body hair and the food that we eat.

Global cities are important sites for the exchange of flows between the global and the body. Global cities have a range of "ethnic" restaurants, stores and shops that allow "exotic" goods and services to be consumed. The cosmopolitan body is more easily sustained and reproduced in the global city. And globalizing cities seek to attract and sustain the cosmopolitan body.

Hybridity, a recognized feature of a postmodern global world, is created and reinforced by the movement of our bodies. Immigrants to richer countries can now more easily connect with families and can more easily retain their joint status. Keeping in touch across international boundaries, literally and by voice and email as well as by money transfers, is an important ingredient of cultural hybridity, economic globalization and cosmopolitanism.

Bodies are both tethered and untethered by global connections. It is easier and cheaper for people to move across the world and yet we are increasingly tied to our means of communication. Access to the Internet and the telephone is now vital to maintaining our connections. We have greater mobility but greater reliance on the grid of international communications. Globalization has both untethered us from the heavier constraints of international travel and has also tethered us to the circuits of telecommunication. Global cities are the points in the world where we can both travel and be connected.

Leonardo had a clear model of body–cosmos connection. The human body was the cosmos in miniature. The cosmos was the human body writ large. The precise connections between cosmos and body are no longer believed. Few now subscribe to the model of the four elements of fire, earth, air and water as fundamental properties of the cosmos. In these more complex times there is no equivalent model. But if there were, the three primary elements would be the global world, the human body and the globalizing cities. Globalization creates, nurtures, sustains and dissolves specific bodily boundaries especially in globalizing cities. Globalization is quite literally embodied and this embodiment is particularly evident in global and globalizing cities.

References

Abramson, D. B. (1997) "Marketization" and institutions in Chinese inner-city redevelopment: a commentary on Lu Junhua's Beijing's Old and Dilapidated Housing Renewal. *Cities* 14(2), 71–75.

Anderson, L. J. (2000) Letter From Havana – The Old Man and the Boy. *The New Yorker*. October 21 and 28, 224–237.

Andranovich, G, M., Burbank, J. and Heying, C. H. (2001) Olympic Cities: lessons learned from mega-event politics. *Journal of Urban Affairs* 23(2), 113–131.

*Anonymous (2003a) Localized effects of globalization: the case of Ciudad Juarez, Chihuahua, Mexico. *Urban Geography*.

*Anonymous (2003b) China's extended metropolitan regions: formation, delimitation and impact on regional development. *Transactions of the Institute of British Geographers*.

*Anonymous (2003c) Transplanting cityscapes: the use of imagined globalization in housing commodification in Beijing. *Area*.

Appadurai, A. (1996) *Modernity at Large: Cultural Dimensions of Globalization*. Minneapolis: University of Minnesota Press.

Ashworth, G. and Voogd, H. (1990) *Selling the City: Marketing Approaches in Public Sector Urban Planning*. London: Belhaven Press.

Bachrach, S. D. (2000) *The Nazi Olympics: Berlin 1936*. Boston: Little Brown and Co.

Bauman, Z. (2000) *Community: Seeking Security in an Insecure World*. Cambridge: Polity Press.

Beaverstock, J. V. (2002) Transnational elites in global cities: British expatriates in Singapore's financial district. *Geoforum* 33(4), 525–538.

Beaverstock, J. V. and Boardwell, J. T. (2000) Negotiating globalization, transnational corporations and global city financial centres in transient migration studies. *Applied Geography* 20(2), 227–304.

Beaverstock, J. V., Smith, R. G. and Taylor, P. J. (1999a) The long arm of the law: London's law firms in a globalizing world-economy. *Environment and Planning A* 31(10): 1857–1876.

Beaverstock, J. V., Smith, R. G. and Taylor, P. J. (1999b) Geographies of globalization: US law firms in world cities. *Urban Geography* 21(2), 95–120.

Beaverstock, J. V., Smith, R. G. and Taylor, P. J. (2000a) World city network: a new metageography? *Annals of the Association of American Geographers* 90(1), 123–135.

Beaverstock, J. V., Smith, R. G., Taylor, P. J., Walker, D. R. F. and Lorimer, H. (2000b) Globalization and world cities: some measurement methodologies. *Applied Geography* 20(1), 43–63.

Behrman, J. and Rondinelli, D. (1992) The cultural imperatives of globalization: urban economic growth in the 21st century. *Economic Development Quarterly* 6(2), 115–126.

Beijing-China (2000) http://www.beijing.gov.cn.

Bell, D. and Valentine, V. (1996) *Consuming Geographies: You Are Where You Eat.* London: Routledge.

Bergesen, A. and Bata, M. (2002) Global and national inequality: are they connected? *Journal of World Systems Research* 8(1), 130–144.

Birmingham, J. (1999) *Leviathan: The Unauthorised Biography of Sydney.* Milson's Point, NSW: Random House.

Borja, J. (1996) The city, democracy and governability: the case of Barcelona. *International Social Science Journal* 48, 85–93.

Bourdieu, P. (1984) *Distinction: A Social Critique of the Judgement of Taste.* London: Routledge.

Brenner, N. (1998) Global cities, glocal states: global city formation and state territorial restructuring in contemporary Europe. *Review of International Political Economy* 5(1), 1–37.

Brenner, N. (1999) Globalisation as reterritorialisation: the re-scaling of urban governance in the European Union. *Urban Studies* 36(3), 431–451.

Brenner, N. (2001) The limits of scale? Methodological reflections on scalar structuration. *Progress in Human Geography* 25(4), 591–614.

Brinkhoff, T. (2001) The principal agglomerations of the world. http://www.citypopulation.de, 10–14–2001.

Caggiano, C. (1999) 'Microsoft – Maybe you've heard of it?'. *Inc.* May 18.

Calvert, J. (2002). How to buy the Olympics. *Observer Sport Monthly* 21, 32–37.

Cannadine, D. (2001) *Ornamentalism: How the British Saw Their Empire.* Harmondsworth: Allen Lane.

Carrel, T. (2000) Beijing: new face for the ancient capital. *National Geographic* 197(3), 116–137.

Cashman, R. and Hughes, A. (eds) (1998) *The Green Games: A Golden Opportunity.* Sydney: Center for Olympic Studies, University of New South Wales.

Castells, M. (1989) *The Informational City.* Oxford: Blackwell.

Castells, M. (2000) *The Rise of the Network Society.* Oxford: Blackwell.

Cheshire, P. (1999) Cities in competition: articulating the gains from integration. *Urban Studies* 36(5–6), 843–864.

Connell, J. (ed.) (2000) *Sydney: The Emergence of a World City.* South Melbourne: Oxford University Press.

Conversi, Daniele (1997) *The Basques, the Catalans and Spain: Alternative Routes to Nationalist Mobilisation.* Reno, NV: University of Nevada Press.

Coyula, M. (1996) The neighborhood as workshop. *Latin American Perspectives* 23(4), 90–103.

DeBord, G. (1994) *The Society of the Spectacle.* New York: Zone Books.

Debray, R. (1995) Remarks on the spectacle. *New Left Review* 214, 34–41.

De Moragas, M. and Botella, M. (eds) (1995) *The Keys to Success: The Social, Sporting, Economic and Communications Impact of '92.* Barcelona: Centre d'Estudias Olimpics I del Esport, Universitat Autonoma de Barcelona.

Dutton, M. (1999) Street scenes of subalternity: China, globalization and rights. *Social Text* 17(3), 63–86.

Economist (1998) Cities and growth: paradise overrun. *The Economist* 347, 33–35.

Economist (2002) Special report on Hong Kong and Shanghai: rivals more than ever. *The Economist*, March 30, 19–21.

Economy (2000) http://www.ci.seattle.wa.us/tda/Dseco.html, 1–25–00.

Elkin, S. L. (1987) *City and Regime in The American Republic*. Chicago: University of Chicago Press.

Espy, R. (1979) *The Politics of The Olympic Games*. Los Angeles and Berkeley: University of California Press.

European City of Culture: Prague 2000 (2000) http://www.czechsite.com/ecc. html, 1.25.00.

Fainstein, S. (2001) Inequality in global city-regions. In A. J. Scott (ed.) *Global City-Regions*. Oxford: Oxford University Press.

Frank, R. H. (2000) *Luxury Fever*. New York: Free Press.

Friedmann, J. (1986) The world city hypothesis. *Development and Change* 17, 69–83.

Friedmann, J. (1995) Where we stand: a decade of world city research. In Paul L. Knox and Peter J. Taylor (eds) *World Cities in a World System*. Cambridge: Cambridge University Press.

Friedmann, J. and Wolff, G. (1982) World city formation: an agenda for research and action. *International Journal of Urban and Regional Research* 3(2), 309–344.

GAWC (Globalization and World Cities) (2002) http://www.lboro.ac.uk/gawc/ publicat.html.

Gitter, Robert and Schueler, Markus (1998) Low unemployment in the Czech Republic: miracle or mirage? *Monthly Labor Review* 121(8), 31–45.

Gollwitzer, H. (1969) *Europe in The Age of Imperialism, 1880–1914*. London: Thames and Hudson.

Goodman, P. S. (2003) White-collar work a booming U.S. export. *Washington Post*, April 2, E1 and E5.

Goodwin, M., Duncan, S. and Halford, S. (1993) Regulation theory, the local state and the transition of urban politics. *Environment and Planning D: Society and Space* 11(1), 67–88.

Gray, M., Golob, E. and Markusen, A. (1996) Big firms, long arms, wide shoulders: the hub-and-spoke industrial district in the Seattle region. *Regional Studies* 30, 651–667.

Grimsley, K. (1999) With jobs plentiful, workers take their pick: South Dakota firms use executive perks to fill basic posts. *Washington Post*, June 5.

Guttman, A. (1992) *The Olympics: A History of the Modern Games*. Urbana, IL: University of Illinois Press.

Hall, P. (1966, reprinted 1984) *The World Cities*. London: Weidenfeld and Nicholson.

Hall, T. and Hubbard, P. (1996) The entrepreneurial city: new urban politics, new urban geographies? *Progress in Human Geography* 20(2), 153–174.

Hall, T. and Hubbard, P. (eds) (1998) *The Entrepreneurial City: Geographies of Politics, Regime and Representation*. New York: Wiley.

Halperin, Maurice (1994) *Return to Havana: The Decline of Cuban Society under Castro*. Nashville: Vanderbilt University Press.

Hanna, M. (1999) *Reconciliation in Olympism: Indigenous Culture in the Sydney Olympiad*. Petersham, NSW: Walla Walla Press.

Hannerz, U. (1993) The cultural roles of world cities. In A. P. Cohen and K. Fukuo (eds) *Humanizing the City*. Edinburgh: Edinburgh University Press.

Hargreaves, J. (1986) *Sport, Power and Culture*. Cambridge: Polity Press.

Hargreaves, J. (2000) *Freedom for Catalonia: Catalan Nationalism, Spanish Identity and the Barcelona Olympic Games*. Cambridge: Cambridge University Press.

Harvey, D. (1989) From managerialism to entrepreneurialism: the transformation in urban governance in late capitalism. *Geografiska Annaler* 71B(1), 3–18.

Harvey, T. (1996) Portland, Oregon: regional city in a global economy. *Urban Geography* 17, 95–114.

Haseler, S. (2000) *The Super-Rich: The Unjust New World of Global Capitalism*. New York: St Martin's Press.

Hessler, P. (2001) Boomtown girl. *The New Yorker*, May 28, 108–119.

Hill, C. (1992) *Olympic Politics*. Manchester: Manchester University Press.

Hindbjorgen, Dan (1999) 'Re: Sioux Falls research.' E-mail to the author, October 18.

Holcomb, B. (1993) Revisioning place: de- and re-constructing the image of the industrial city. In G. Kearns and C. Philo (eds) *Selling Places: The City as Cultural Capital, Past and Present*. Oxford: Pergamon Press.

Howlett, D. (1998) Sioux Falls is booming – reluctantly. *USA Today*, July 23.

Hughes, Robert. (1992) *Barcelona*. New York: Knopf.

Hutton, W. (1995) *The State We're In*. London: Jonathan Cape.

IOC (2002) http://www.olmpic.org/ioc/e/org/ioc_members_e.html.

Jennings, A. (1996) *The New Lords of the Rings*. New York: Simon and Schuster.

Jennings, A. and Sambrook, C. (2000) *The Great Olympic Swindle*. New York: Simon and Schuster.

Jonas, A. and Wilson, D. (eds) (1999) *The Urban Growth Machine: Critical Perspectives Two Decades Later*. Albany: State University of New York Press.

Kaplowitz, Donna Rich (1998) *Anatomy of a Failed Embargo: U.S. Sanctions Against Cuba*. Boulder, CO: Lynne Rienner Publishers Inc.

Kaye, L. (1992) Pride and publicity: Peking's Olympics bid could be aimed at Taiwan. *Far Eastern Economic Review*, August 13.

Kearns, G. and Philo, C. (eds) (1993) *Selling Places: The City as Cultural Capital, Past and Present*. Oxford: Pergamon Press.

Keil, R. (1998) *Los Angeles: Globalization, Urbanization and Social Struggles*. New York: John Wiley.

Keil, R., Wekerle, G. and Bell, D. (eds) (1996) *Local Places in The Age of The Global City*. Montreal: Black Rose Books.

King, A. (2002) Speaking from the margins: postmodernism, transnationlism and the imagining of contemporary Indian urbanity. In R. Grant and J. R. Short (eds) *Globalization and the Margins*. Basingstoke and New York: Palgrave.

Kong, L. and Yeoh, B. S. A. (2003) *The Politics of Landscape in Singapore*. Syracuse: Syracuse University Press.

Korporaal, G. and Evans, M. (1999) Games people play. *Sydney Morning Herald*, February 11: 2.

Krätke, S. and Schmoll, F. (1991)The local state and social restructuring. *International Journal of Urban and Regional Research* 15(1), 542–552.

Kristof, N. and Wudunn, S. (1995) *China Wakes: The Struggle for the Soul of a Rising Power*. New York: Vintage Books.

Krugman, P. (1991) *Geography and Trade*. Cambridge: MIT Press.

Lauria, M. (1997) *Reconstructing Urban Regime Theory: Regulating Urban Politics in a Global Economy*. London: Sage.

Leaf, M. (1995) Inner city redevelopment in China: implications for the city of Beijing. *Cities* 12(3), 149–162.

Lenskyj, H. J. (2000) *Inside The Olympic Industry*. Albany: State University of New York Press.

Leo, C. (1997) City politics in an era of globalization. In M. Lauria (ed.) *Reconstructing Urban Regime Theory: Regulating Urban Politics in a Global Economy*. London: Sage.

Lowndes, V. (1995) Citizenship and urban politics. In D. Judge, G. Stoker and H. Wolman (eds) *Theories of Urban Politics*. London: Sage.

Lundberg, F. (1988) *The Rich and the Super-Rich*. New York: Citadel Press.

MacCurdy, E. (ed.) (1956) *The Notebooks of Leonardo da Vinci, volume 1*. New York: George Braziller.

Magdalinski, T. (2000) The reinvention of Australia for the Sydney Olympic Games. *International Journal of the History of Sport* 17, 305–322.

Maguire, J. (1994) Sport, identity politics, and globalization: diminishing contrasts and increasing varieties. *Sociology of Sport Journal* 11, 398–427.

Mandell, R. D. (1971) *The Nazi Olympics*. London: Souvenir Press.

Marshall, A. (1922) *The Principles of Economics*. London: Macmillan.

McCollum, R. H. and McCollum, D. F. (1981) Analysis of ABC TV Coverage of the 21st Olympiad Games. In J. Segrave and D. Chu (eds) *Olympism*. Champaign, IL: Human Kinetics.

McDonogh, G. W. (1987) The geography of evil: Barcelona's barrio chino. *Anthropological Quarterly* 60, 174–184.

McKay, M. and Plumb, C. (2001) *Reaching Beyond The Gold*. London: LaSalle Investment Management.

Merwin, J. (1983) Citibank put little South Dakota back on the map. *Forbes*, November 21.

Mitchell, D. (2003) *The Right to The City*. New York and London: Guilford.

Molotch, H. (1993) The political economy of growth machines. *Journal of Urban Affairs* 15, 29–53.

Moorhouse, G. (1999) *Sydney*. St Leonards, NSW: Allen and Unwin.

Morris, J. (1992) *Sydney*. London: Viking.

Ohmae, K. (1995) *The End of the Nation State and the Rise of Regional Economies*. New York: Free Press.

Orwell, George (1980) *Homage to Catalonia*. San Diego: Harvest/HBJ.

Pahl, R. (2001) Market success and social cohesion. *International Journal of Urban and Regional Research* 25(4), 879–883.

Parnreiter, C. (2002) Mexico: the making of a global city. In Saskia Sassen (ed.) *Global Networks, Linked Cities*. New York and London: Routledge.

Peck, J. and Tickell, A. (1995) Business goes local: dissecting the "business agenda" in Manchester. *International Journal of Urban and Regional Research* 19(1), 55–78.

Plaza, B. (2000) Evaluating the influence of a large cultural artifact in the attraction of tourism: the Guggeneheim Museum Bilbao case. *Urban Affairs Review* 36(2), 264–274.

Pollock, J. (1997) City characteristics and coverage of China's bid to host the Olympics. *Newspaper Research Journal*, Summer/Fall.

Pomfret, J. (1999) China's art and soul: Beijing and Shanghai try for the cultural capital title. *Washington Post*, February 2.

Preeg, Ernest H. (1994) Cuba and the Caribbean. *The Conference Board Global Business White Papers* 9.

Preeg, Ernest H. and Levine, Jonathan D (1993) *Cuba and the New Caribbean Economic Order*. Washington, DC: The Center for Strategic and International Studies.

Preuss, H. (2000) *Economics of The Olympic Games*. Sydney: Walla Walla Press.

Purcell, M. (2002) Politics in global cities: Los Angeles charter reform and the new social movement. *Environment and Planning A* 34(1): 23–42.

Richter, A. (1952) *Selections from the Notebooks of Leonardo da Vinci*. Oxford: Oxford University Press.

Ritter, Archibald R. M. and Kirk, John M. (eds) (1995) *Cuba in the International System: Normalization and Integration*. New York: St Martin's Press.

Roche, M. (1992) Mega-events and micro-modernization; on the sociology of the new urban tourism. *British Journal of Sociology* 43, 564–600.

Rosenfeld, J. (2000) Experience the real thing. *Fast Company*, January/February, 184–196.

Rupert, M. (2000) *Ideologies of Globalization*. London: Routledge.

Sanchez, J. (1992) Societal responses to changes in the production system: the case of Barcelona metropolitan region. *Urban Studies* 29(6), 949–964.

Sassen, S. (1991) *The Global City: New York, London, Tokyo*. Princeton: Princeton University Press.

Sassen, S. (1994) *Cities in a World Economy*. Thousand Oaks, CA: Pine Forge Press.

Sassen, S. (1999) Embedding the global in the national: implications for the role of the state. In David A. Smith, Dorothy J. Solinger and Steven C. Topik (eds) *States and Sovereignty in the Global Economy*. London and New York: Routledge.

Sassen, S. (2000) The global city: strategic site/new frontier. In E. Isin (ed.) *Democracy, Citizenship and the Global City*. New York: Routledge.

Schor, J. (1992) *The Overworked American*. New York: Basic Books.

Schor, J. (1998) *The Overspent American*. New York: Basic Books.

Schwab, P. (1999) *Cuba: Confronting the U.S. Embargo*. New York: St Martin's Press.

Scott, A. J. (2000) *The Cultural Economy of Cities*. London: Sage.

Scott, A, J. (ed.) (2001) *Global City-Regions*. Oxford: Oxford University Press.

Segre, Roberto, Coyula, Mario and Scarpaci, Joseph L. (1997) *Havana: Two Faces of the Antillean Metropolis*. Chichester: John Wiley & Sons.

Short, J. R. (1999) Urban imaginers; boosterism and the representation of cities. In A. Jonas and D. Wilson (eds) *The Urban Growth Machine: Critical Perspectives Two Decades Later*. Albany: State University of New York Press.

Short, J. R. (2001) *Global Dimensions: Space, Place and The Contemporary World*. London: Reaktion.

Short, J. R. and Kim, Y. (1998) Urban crises/urban representations: selling the city in difficult times. In Tim Hall and Phil Hubbard (eds) *The Entrepreneurial City*. Chichester: Wiley.

Short, J. R. and Kim, Y.-H. (1999) *Globalization and the City*. Harlow: Longmans.

Short, J. R., Benton, L., Luce, B. and Walton, J. (1993) The reconstruction of the image of a postindustrial city. *Annals of Association of American Geographers* 83(2), 207–224.

Short, J. R., Kim, Y., Kuus, M. and Wells, H. (1996) The dirty little secret of world cities research. *International Journal of Urban and Regional Research* 20, 697–715.

Simpson, Fiona (1999) Tourist impact in the historic centre of Prague: resident and visitor perceptions of the historic built environment. *The Geographical Journal* 165, 173–185.

"Sioux Falls Designated for Port of Entry" (1995) *Aberdeen American News*, October 19, A4.

Sioux Falls Development Foundation (1999) *Directions* 23(5), September – October.

'Sioux Falls No. 1 Place to Live' (1992) *Christian Science Monitor*, October 4.

Sklair, L. (1997) The transnational capitalist class. In J. G. Carrier and D. Miller (eds) *Virtualism: A New Political Economy*. Oxford: Berg.

Smith, N. (1996) *The New Urban Frontier: Gentrification and The Revanchist City*. New York: Routledge.

Smith, R. C. (2001) *The Wealth Creators: The Rise of Today's Rich and Super-Rich*. New York: Truman Talley Books.

Spearritt, P. (1999) *Sydney's Century: A History*. Sydney: University of New South Wales Press.

Stoker, G. (1990) Regulation theory, local government and the transition from Fordism. In D. S. King and J. Pierre (eds) *Challenges to Local Government*, London: Sage.

Stone, C. N. (1989) *Regime Politics: Governing Atlanta, 1946–1988*. Kansas: University of Kansas Press.

Storper, M. (1997) *The Regional World*. New York: Guildford.

Strenk, A. (1981) Amateurism: myth and reality. In J. Segrave and D. Chu (eds) *Olympism*. Champaign, IL: Human Kinetics.

Strout, E. (1999) Sioux Falls, South Dakota. *Sales and Marketing Management*, April.

Swyngedouw, E. (2000) Authoritarian governance, power and the politics of rescaling. *Environment and Planning D: Society and Space* 18, 63–76.

Sykora, L. (1994) Local urban restructuring as a mirror of globalisation process: Prague in the 1990s. *Urban Studies* 31, 1149–1166.

Taylor, P. J. (1999) Worlds of large cities; pondering Castells' space of flows. *Third World Planning Review* 21(3), iii–x.

Taylor, P. J. (2000) World cities and territorial states under conditions of contemporary globalization. *Political Geography* 19(1), 5–32.

Taylor, P. J. (2001) Specification of the world city network. *Geographical Analysis* 33(2), 181–194.

Taylor, P. J., Catalano, G. and Walker, D. R. (2001) Measurement of the world city network. *Research Bulletin 43* (http:www.lboro.ac.uk/gawc/rb/rb43.html).

Thomas, B. (1990) Catalonia: the factory of Spain (special advertising supplement). *Forbes* 145(12), A1–A3.

Thompson, G. (2001) Fallout of U.S. recession drifts south into Mexico. *The New York Times, Business Section*, December 26, 1–2.

Thrift, N. (1994) On the social and cultural determinants of international financial centres: the case of the City of London. In S. Corbridge, N. Thrift and R. Martin (eds) *Money, Power and Space*. Oxford: Blackwell.

Thrift, N. (2000) Performing cultures in the new economy. *Annals of the Association of American Geographers* 90(4), 674–692.

Thrift, N. (2001) 'It's the romance, not the finance, that makes the business worth pursuing': disclosing a new market culture. *Economy and Society* 30(4), 412–432.

Toensmeier, P. (1993) Editorial. *Modern Plastics* 70(9), 98–107.

Tomlinson, A. (1996) Olympic spectacle: opening ceremonies and some paradoxes of globalization. *Media, Culture and Society* 18, 583–602.

Tomlinson, A. and Whannel, G. (eds) (1984) *Five-Ring Circus: Money, Power and Politics at the Olympic Games*. London: Pluto Press.

Turnbull, L. (1999) *Sydney: Biography of a City*. Milson's Point, NSW: Random House.

UNDP (2002) UNU/WIDER – World Income Inequality Database. New York: United Nations.

Urry, J. (2001) *Mobilities*. London: Routledge.

Veblen, T. (1985) *A Theory of the Leisure Class*. London: Allen and Unwin.

Ward, S. V. (1998) *Selling Places: The Marketing and Promotion of Towns and Cities, 1850–2000*. London: Spon Press.

Watson, S. and Murphy, P. (1997) *Surface City: Sydney at the Millennium*. Annandale, NSW: Pluto Press.

Wilkinson, T. (1996) A city of the '90s with Leave it to Beaver values. *Christian Science Monitor*, November 29.

Williams, A. R. (1999) The rebirth of Old Havana. *National Geographic* 195(6), 36–45.

Williams, R. (1976) *Keywords: A Vocabulary of Culture and Society*. London: Fontana.

Wilson, H. (1996) What is an Olympic city? Visions of Sydney 2000. *Media, Culture and Society* 18, 603–618.

World Bank (2001) http://www.worldbank.org/data/countrydata/countrydata.html.

World Markets Country Analysis (2002) http://www.worldmarketsanalysis.com.

Wu, F. (2001a) Housing provision under globalization: a case study of Shanghai. *Environment and Planning A* 33, 1741–1764.

Wu, F. (2001b) China's recent urban development in the process of land and housing marketisation and economic globalization. *Habitat International* 25, 273–289.

Wu, F. (2000c) Place promotion in Shanghai, PRC. *Cities* 17, 349–361.

Wu, F. (2000d) The global and local dimensions of place-making: remaking Shanghai as a world city. *Urban Studies* 37, 1359–1377.

Zelinsky, W. (1991) The twinning of the world cities in geographic and historical perspective. *Annals of the Association of American Geographers* (81), 1–31.

* I regularly receive manuscripts from academic journals to review. In these three cases I have cited them as I have drawn on some of their ideas or presented their data. I do not know who the authors are but wanted to acknowledge their work.

Index